ALCOHOL;

AS A

FOOD AND MEDICINE.

A PAPER FROM THE TRANSACTIONS

OF THE

INTERNATIONAL MEDICAL CONGRESS,

AT PHILADELPHIA, SEPTEMBER, 1876.

BY

EZRA M. HUNT, A.M., M.D.,

PRESIDENT OF THE SECTION OF THE AMERICAN MEDICAL ASSOCIATION ON STATE
MEDICINE AND PUBLIC HYGIENE; VICE-PRESIDENT OF THE AMERICAN PUBLIC
HEALTH ASSOCIATION; DELEGATE TO THE INTERNATIONAL MEDICAL
CONGRESS FROM THE NEW JERSEY STATE MEDICAL SOCIETY, ETC.

NEW YORK:

National Temperance Society and Publication House,

58 READE STREET.

1882.

PREFACE.

THE International Medical Congress, which met in Philadelphia, in September, 1876, and was the most representative medical body ever convened in this country, consisted of over 600 delegates.

The National Temperance Society, feeling the importance of some declaration from such an assembly as to the status of Alcohol as an alleged food or a medicine, addressed to it the following memorial in relation thereto :

TO THE INTERNATIONAL MEDICAL CONGRESS, PHILADEL-PHIA :

The National Temperance Society sends greeting, and respectfully invites from your distinguished body a public declaration to the effect that alcohol should be classed with other powerful drugs ; that when prescribed medicinally, it should be with conscientious caution and a sense of grave responsibility ; that it is in no sense food to the human system ; that its improper use is productive of a .large amount of physical disease, tending to deteriorate the human race ; and to recommend, as representatives of enlightened science, to your several nationalities, total abstinence from alcoholic beverages.

W. E. DODGE, *President.*

J. N. STEARNS, *Corresponding Secretary.*

NEW YORK, *September* 5, 1876.

The Woman's National Christian Temperance Union and the New York Friends Temperance Union also presented kindred memorials.

Preface.

The Congress was divided into sections for the more thorough discussion of the various medical topics. "The Programme of Public Business" announced a paper on "Alcohol in its Therapeutic Relation as a Food and a Medicine" before the "Section on Medicine." The memorials were referred to that Section for consideration, after the reading of the paper on Alcohol, and for a subsequent report to the Congress. The discussion was one of marked ability and earnestness. After various propositions and resolutions had been offered, the conclusions which the author of this paper presented were moved and quite unanimously adopted as the sentiments of the Section on Medicine. ·As such they were reported for acceptance to the General Congress, and by it ordered to be transmitted as a reply to the memorialists. This paper thus became so relevant to the action of the body in this regard that we have secured it for publication in full. Although so strictly medical, its important bearing on Temperance is obvious. The author·has preferred to leave the paper as presented to the Congress, with only an occasional note or reference.

THE THERAPEUTIC VALUE

OF

ALCOHOL AS FOOD,

AND AS A

MEDICINE.

BY EZRA M. HUNT, M.D.

WE think it will readily be granted by the members of this Medical Congress that no article named in the Materia Medica bears more important relations to practical medicine than that which is borne by alcohol. In China or Japan, opium might contend for the palm of prominence, but in a land in which spirituous liquors, in some form or other, are prescribed by thousands of physicians, and used daily as an assumed health-beverage by hundreds of thousands of the people, the bearing of such use on medical practice is of indisputable importance. Alcohol stands preëminent as the one article of Materia Medica more familiar to the public mind and the public stomach than any other.

If there is one serious and direct question which, beyond all others, humanity has a right to ask of the medical profession, it is, What are the claims of alcoholic liquors in their relations to health and disease?

While the materia medicist, the moralist, and the

writer on practical medicine may each render prominent the stand-point from which their views are taken,
all alike agree that, practically, alcohol is a great in-
cubus upon the public health.

The sanitary publicist from his own observation
readily accords in the view, and all unite to ask the
needs of use, the limits to abuse, and our professional
and personal relations thereto.

Questions of surgery, obstetrics, and the like, have
mostly to do with points of contact between the physician and the individual patient, and, therefore, have
a strictly clinical bearing. Not so with alcohol. There
are some physician's problems that are to be worked
out for all humanity. We can not be true to science,
and much less to the *morale* of our profession, without meeting questions as to our air, our foods, our
drinks, our medicines, with all their bearings upon
health as they concern the masses.

We could not even draw back from this as a social
and moral question. The citizen, when he comes to
be the physician, does not cease to be the man.

Rather his manhood lays all its opulence at the
door of his profession, and bids him, as far as possible, practice and teach in the direction of all physical,
social, and moral welfare, if he would conserve the
true spirit of his calling.

Indeed, could we meet a man with soul so dead as
that he should be able to infuse the stoicism of Cle-
anthes into his own theoretical pursuit of medical
science, we should still stand before him with the
material nature, and, in justice to the body alone,
plead that he should define the status of alcohol as a

food, its therapy as a medicine, and then indicate to us, with all the strictness of rigid analysis and all the exactness of classified experience, its proper conservatism in the interests of health.

Our duty in this regard is all the more imminent and vital because its use so often involves its abuse, and because the result is so manifestly inimical to human health. What consumption is among diseases, that is alcohol among medicines. How to limit the causes of the one is not more a professional question with us than how to abate the necessity for the effects of the other. It can not be concealed that the habit of the employment of alcoholics as a beverage is supported by the persuasion that they have an important value as a food, or are reparative and recuperative, and so medicinal. Thus, as half food and half medicine, they are accredited into popular use. If they were not already before us in our therapeutics, this very fact would so thrust them into professional contact that it would be professional pusillanimity not to meet the issues involved.

They are before us in honest fealty to our profession, not less than in that grand fealty which belongs to relative and regulative truth, and which it is the blessed meed of our science and our art ever to recognize and accord.

We propose to consider this subject by inquiring—

1. What is the value of alcohol as a food?

2. What is its value as a medicine?

3. How far it is modified by the variable compositions of spirituous liquors, or by unreliability in manufacture.

Any article to rank as a food must be convertible into tissue or force, in such a way as to contribute to healthy vitality, and aid the body in the performance of its normal functions. This includes that energy which is needed in the execution of its own processes of nutrition and repair, and that which must be generated to fit it for the expenditure of proper force in its contact with the world about it.

So definite is the relation between the human system and the usual foods by which it is sustained and propelled, that in respect to most of them we are not left in doubt. If we take any one of the indisputable aliments and subject it to chemical analysis, and then apply the same process of examination to human material, we are not slow to trace the correspondence of the two.

Since chemistry has come to be more perfect in its methods of analysis, and physiological investigators have been careful to study relations and test them by science and art combined, these adaptations are seen to be systematic and definite.

If, for instance, we commence with milk, which nature so plainly indicates as the one food for the child in its earliest life, we find in it the very classes of foods which chemistry and physiology show to be in demand for vitality.

We are able to discern very distinctly in this fluid four leading divisions of aliments, viz., (*a*) albuminoid substances, (*b*) fatty material, (*c*) sugar, and (*d*) water, with various salts in solution.

On the other hand, the chemico-physiologist, looking into the human organism, and inquiring what are its demands for sustenance and force, finds nature there pointing with definiteness to these classes of foods as all in demand for its supply.

His only possible puzzle might be in the want of albumen in form, and in the absence of the starches which are found afterward to have so prominent a bearing on sustentation or the exercise of force. As the transition to the use of these is so evidently in the order of nature, and so soon takes place in the case of the child, he might well be led to be inquisitive of science in this direction, and so seek to know whether there was not a similarity or identity of composition after all. A few experiments reveal to him an identity closer than could have been predicated.

The milk analysis shows itself as the reliable key to the study of the other foods.

The casein, for instance, of the milk, which at first seemed distinct from the albumen and fibrin of human flesh, is found to be so far identical that J. Alfred Wanklyn, in his recent analyses, says: "The ultimate composition of casein is not distinguishable from that of albumen and fibrin;" so that "under the name of casein it will be convenient to designate the entire nitrogenous constituents of milk, just as under the name of gluten the entire nitrogenous portion of flour is comprehended" (see Wanklyn's Milk Analysis, 1876, p. 10).

Attfield in his Chemistry (1873), after speaking of the albumen of eggs, the casein of milk, the fibrin of meat, the gluten or "vegetable fibrin, casein,' etc., of breadstuffs (p. 356), and the vegetable albu

1*

men and casein of plant-juices, seeds, and vegetables, not only notices their identity of element, but says these " albuminoid substances are nearly identical in percentage composition " (p. 378). (See, also, Fowne's Chemistry, p. 823).

Besides, chemistry instantly detects the intimate correspondence of starchy and saccharine substances, so that (Attfield, p. 355) they are classed together in their ultimate results as factors of bodily material and force, not only representing the sugar of milk, but also its fat, since, " by the separation of carbon and oxygen," it is accumulated therefrom (see Fowne, p. 825). If, again, we compare meats and vegetables, we find not only a recurrence of the material in milk, but that in many regards the animal is but the di-gested vegetable. Says the chemist Fowne : " The striking contrast which at first appears in the nature of the food of the two great classes of animals—the vegetable feeders and the carnivorous races—dimin-ishes greatly on close examination ; it will be seen that, so far as the materials of blood, or, in other words, those devoted to the repair and sustenance of the body itself, are concerned, the process is the same." " The food employed for nourishment of the body must have the same, or nearly the same, chemical composition as the body itself, and this is really ful-filled in the case of animals that live exclusively on vegetable substances " (Fowne). " Vegetable albumen, fibrin, and casein are scarcely to be distinguished from the bodies of the same name extracted from blood and milk " (see Fowne and Bridges' Chemistry, 1875, p. 823). " It is also noticed that albumen quite similar to that of eggs is extracted from potatoes and the juices of many succulent plants, and that peas and beans and many of the oily seeds contain a principle which bears the most striking resemblance to the casein of milk " (p. 824).

If we look to the fourth class, containing water and salts in solution, we find these provided in great similarity in the liquids, fleshes, grains, vegetables, seeds, and fruits which help to form the aliment of mankind.

A comparison of all these with the human system enables us to express still more briefly the classification of foods, and to speak of the first as nitrogenous or albuminoid foods, whether found in milk, vegetable, or animal, and of the other three classes of solids as carbonaceous food, the water in all cases serving as the dissolving and diffusing menstruum by which the interchanges of the system are conducted.

We are so far able to trace the nitrogenous foods in the formation of tissue, that they are frequently spoken of as " the plastic elements of nutrition " (see Liebig's Animal Chemistry, Attfield, etc.), because they are traceable as directly assimilated, and the mode of their sustenance is recognizable.

We find, also, that the carbonaceous foods have a different association—are absorbed into the system and disappear. When we study the phenomena of heat and force, and find that quantities of carbon and hydrogen are daily oxidized in the human system, and that the laws of animal heat and of the production of force in the outer world are in the closest correlation, we are not left in doubt that " the tempera⸱

ture of the body must be the result of this exertion of chemical force." Hence, these hydrates of carbon (F.) are often called respiratory foods, " because more immediately concerned in keeping up the temperature of the body by the combustion going on between them and their products and the oxygen of the air in the blood " (Attfield, p. 398).

The distinctness of these groups of foods, and their relations to the tissue-producing and heat-evolving capacities of man, are so definite and so confirmed by experiments on animals and by manifold tests of scientific, physiological, and clinical experience, that no attempt to discard the classification has prevailed.

To draw so straight a line of demarcation as to limit the one entirely to tissue or cell production, and the other to heat and force production through ordinary combustion, and to deny any power of interchangeability under special demands or amid defective supply of one variety, is, indeed, untenable.

This does not in the least invalidate the fact that we are able to use these as ascertained landmarks, just as the terms basic and acid oxides are used in chemistry (F., p. 133), and, with these in view, to designate actual distinctions and the relative values of various foods.

Fowne, for instance, would not draw the line between nitrogenous and carbonaceous foods so closely, as to say that the one is nutritive and the other respiratory, in just the way Liebig indicated ; but yet in reference to the former class, makes the axiom that the portion of the food which is destined to the repair and renewal of the frame itself, consists of substances identical in composition with the body it is to nourish, or requiring but little chemical change to become so. Where, as in the case of the production of sugar by the liver (see Final Causes, *Pr. Rev.*, p. 294, 1876, and F., p. 814), or that of fat in the body, the article is furnished by chemical processes in the system, they are of the nature of well-recognized foods, and their mode of production is ascertainable.

So the carbonaceous aliments, such as the hydro-carbons or cellulose gums, starch, sugar, and fats have such relations to respiration and to the chemical changes occurrent therewith, that he says: "Whatever may be the difficulties attending the investigation of these subjects, one thing is clear, namely: that quantities of hydrogen and carbon are daily oxidized in the body by the free oxygen of the atmosphere" (p. 820.) The combustion takes place not, as was once supposed, in the lungs alone, but in the capillaries of the tissues, with carbon and hydrogen as the ever present factors, with heat as a result, and with the expulsion of water and carbonic acid.

This method of the production of heat and the provision of force, corresponds with what we see outside of the human system, and the experiments of Liebig and Boussingault, and those quite recent of Pettenkoffer and Voit, show that the relations of these articles to temperature and to the expenditure of vital force are quite definite enough to sustain the accuracy of the distinctions.

No one can review the various experiments of Voit and Bischoff, Liebig, Wilson, Playfair, Lan-

kester, Letheby, Frankland, Edward Smith, Haugh-
ton, Moleschott, Cameron, Mopather, Pettenkoffer,
Ranke, Parkes, Wollowicz, Richardson, and many
others, without being satisfied that we are in posses-
sion of indications as to what constitutes a food which
have the character of scientific facts and are sustained
by the comparisons of close experience.

If we seek to know how meat, milk, eggs, grains,
vegetables, and seeds are foods, we are able to tell
how they are assimilated and how in their combustion they generate force. In our fruits we are able to
find sugars, salts, and acids ; to see their modes of
disintegration, and in their forms of lactates, tartrates,
citrates, malates, acetates, trace them to their destina-
tions and recognize their appropriation in the system.

Water, as the grand vehicle of osmotic diffusion
through membranes and amid tissue, as the convey-
ancer of life forces, and, in turn, the remover of used-
up material, is easily traced in its irrigation, its trans-
portation, and its changes, and leaves us in no doubt
as to its ever-flowing beneficence.

With such an outline before us as to the facts of
foods, we come very legitimately to ask IN WHAT RE-
SPECTS ALCOHOL VINDICATES ITSELF AS A FOOD.

Our first search may be for its place amid the nitrogenous, albumenate, or cell-forming foods.

We are at once met by the fact that no histologist
or analytical chemist has ever yet been able to venture
the opinion that tissue nutrition can occur without
the intervention of nitrogen. It must at least exist
in the plasm, from which structure takes place. Nay
more, we are able to trace the direct relation of nitro-

gen to the formation of tissue, and to detect the need of nitrogenous foods for its construction.

There has been such unanimity of consent among those of divergent views in other regards, that alcohol is not a tissue-building food, that it is by quite common consent excluded from this class.

We have never seen but a single suggestion that it could so act, and this a promiscuous guess. One writer (Hammond) thinks it possible that it may " somehow " enter into combination with the products of decay in tissues, and " under certain circumstances might yield *their* nitrogen to the construction of new tissue." No parallel in organic chemistry, nor any evidence in animal chemistry, can be found to surround this guess with the areola of a possible hypothesis.

It has been conclusively proved, says Lionel Beale (*Med. Times*, 1872), that alcohol is not a food, and does not directly nourish the tissues.

" There is nothing in alcohol with which any part of the body can be nourished " (Cameron, Manual of Hygiene, p. 282).

" It is not demonstrable at present that alcohol undergoes conversion into tissue " (Hammond, *Tribune* Lecture, May, 1874).

" Alcohol contains no nitrogen ; it has none of the qualities of the structure-building foods ; it is incapable of being transformed into any of them ; it is, therefore, not a food in the sense of its being a constructive agent in the building up of the body " (Richardson on Alcohol, p. 21).

The food tables of Letheby, Frankland, E. Smith

etc., give no place to alcohol as a tissue-forming food, and only allude to ale and porter as containing some nutritious elements from the presence of other substances.

The period of youth is the one in which it should be used to give muscular development and structure building power, if possessing them ; but authors and experimenters, with one accord, exclude it from the diet of children.

In comparative physiology we find the structure of animals and the laws of growth, repair, and tissue-accretion, quite similar to those in man. Yet none of the careful experiments as to foods or the development of the animal kingdom, has ever given alcohol a place for the construction of tissue.

This undoubted exclusion of alcohol from nitrogenous foods is all the more significant because recent investigations have shown that this class of foods not only make cells, but that it has very important correlation in heat-producing processes.

Not detecting in this substance any tissue-making ingredients, nor in its breaking up any combinations such as we are able to trace in the cell-foods, nor any evidence either in the experience of physiologists or the trials of alimentarians, it is not wonderful that in it we should find neither the expectancy or the realization of constructive power.

II.

Our next search for it, therefore, must be amid the carbonaceous or heat and force evolving foods.

Liebig, in his classification, placed it the last in this list, with an interrogation mark by its side.

That question has since been asked in many ways and with many experiments.

The first usual test for a force-producing food, and that to which other foods of that class respond; is the production of heat in the combination of oxygen therewith. This heat means vital force, and is in no small degree a measure of the comparative value of the so-called respiratory foods.

The close studies of modern investigation have shown how uniform is the law of the evolution of energy through the processes by which heat is produced. Heat as a mode of motion is the all-prevalent law in the animal organism as well as in the world of its contact.

If we examine the fats, the starches, and the sugars, we can trace and estimate the processes by which they evolve heat and are changed into vital force, and can weigh the capacities of different foods.

We find that the consumption of carbon by union with oxygen is the law, that heat is the product, and that the legitimate result is force, while the result of the union of the hydrogen of the foods with oxygen is water. If alcohol comes at all under this class of foods, we rightly expect to find some of the evidences which attach to the hydro-carbons.

For a long time on this point, so-called experience was assumed not to need the confirmation of chemico-physiological facts; for it was accepted as a fact that heat must be, and was, an effect from alcohol introduced into the body, and, therefore, it must belong to the class of heat-producing foods.

The sensation of heat which is experienced when alcohol is imbibed, naturally sustained this view, and popular and medical belief accepted it as a grand calorifacient both in health and disease. It was thus certified to be a self-evident warmer.

But as the physiology of digestion came to be more thoroughly understood, and especially that of the nervous system, we found that even this effect as a local stimulus, or, perhaps, as a local irritant, did not suffice for proof. As even Hammond expresses it, " It is one of those abnormal manifestations of nerve-action met with in several other conditions of the system." At least it is a pungency quite different from the heat evolved by the chemical transformation of a hydro-carbon, or such as results in physiological force.

The idea that it is a heat-producing food, and so, like calorific foods, is consumed in the system according to the usual law of disintegration, was so far accepted that Anstie, speaking of its *anti-pyretic* or cooling effect as now admitted, says : " Until within the last few years it was never even suspected " (see Practitioner, Vol. III., 1873, p. 422). In his advocacy, while claiming that somehow alcohol must be oxidized and converted into force, he then says, " this force can not be heat."

In other words, the most remarkable *exception* in all the laws of *energy* must be hypothecated in order to make tenable the oxidation of alcohol, so as somehow to elaborate life-force. Yet, as H. R. Wood, Jr.,

in his recent Materia Medica, expresses it, " No one
has been able to detect in the blood any of the ordinary products of its oxidation."

Some of the most thorough experiments in animal
chemistry have been in reference to this combustion,
as a result of which carbonic acid (carbon dioxide)
and water result, and heat or animal temperature is
evolved.

The process is quite analogous to that occurrent in
the outer world, when heat and force are produced.
So declarative is this that in the apparatus of Pettenkoffer " the quantity of carbon and of hydrogen
in a stearine candle burnt in the apparatus could be
determined as accurately by the quantity of carbon
dioxide and water produced, as by an organic analysis " (F., p. 822).

The calculation in the human organism is all so
much in the same line of direction, that a distinguished chemist (Fowne, p. 825) has said : " The
chemical phenomena observed in the animal system
resemble so far those produced out of the body by
artificial means, that they are all, or nearly all, so far
as known, changes in a descending series;" *i. e.*,
changes in which energy is evolved by disintegration.
In the case of the so-called respiratory foods the
" combustible contains carbon and hydrogen," and
" part or all of the high temperature of the body
must be the result of this exertion of chemical force."
These are not opinions, but chemical facts.

In the case of such a body as alcohol, if it contrib-
utes to such a result or undergoes this force-producing
evolution, a prominent and pronounced indication

ought to be found in an increase of internal temperature under free administration.

But the facts elicited on this point are, of late, so strongly in the other direction, that the time-honored advocacy of alcohol as a means of sustaining animal temperature has quite subsided with all those familiar with the evidence. The pathognomonic evidence of food power has vanished.

Since the introduction of the thermometer as a test of internal temperature and its carefully registered results, we are able to estimate with much accuracy the production of heat by various foods or amid disturbing influences.

"In the very first stage of the imbibition of alcohol," says Richardson, "there is found a slight increase of temperature now and then, in a confirmed inebriate, amounting to one and a half degrees, but this is brief, and is itself the result of a cooling process." The same occurs when a cold bath is administered (see Prof. Wm. H. Thompson, *Medical Times*, Oct., 1875). It is the shock or temporary physical effect from the nervous relaxation of the circulatory system, and so "an unfolding of the larger sheet of the warm blood and a quicker radiation of heat from a large surface" (R., p. 70).

A similar result takes place in blushing, from a temporary relaxation and suffusion of the capillaries, because of the temporary modification of the contractile power of the vaso-motor nerves which are supplied to the circulatory tubing.

When Todd, in advance of experimental evidence, announced the observation that alcohol was a reducer of temperature in fevers, most regarded the proposition as erratic, because contrary to experience, and so to the *communis sensus* of practitioners. Now so uniform is the testimony of the most of German, French, English, and American physicians as to the unavailability of alcohol as a heat-producer, that its anti-pyretic power is its boasted value as a medicine by many. H. R. Wood, Jr., in his Materia Medica, 1876, says the accord is such " that it does not seem worth while to occupy space with a discussion of the subject."

Liebermeister, one of the most learned contributors to Ziemssen's Cyclopædia of the Practice of Medicine, 1875, says : " I long since convinced myself by direct experiments that alcohol, even in comparatively large doses, does not elevate the temperature of the body in either well or sick people " (Vol. 1, p. 223).

Indeed, outside experience had almost forestalled the views of physicists and clinicians. Says Wood : "At a time when physiologists believed that alcohol had a heat-producing power, the Northern navigators had learned that the free use of spirits, far from enabling a man to withstand habitual exposure to intense cold, very naturally lessened his powers of resistance." "In the Arctic regions," says Edward Smith, "it was proved that the entire exclusion of spirits was necessary in order to retain heat under those extremely unfavorable conditions." In the voyage of search after Sir John Franklin, no alcoholic stimulus was used, and amid the cold whaling grounds of the North the needlessness of alcohol has been well substantiated (see Parkes, Hooker, etc.)

This fact, then, that alcohol is shown by the direct test of experiment, confirmed by experience, not to be a sustainer of animal heat, goes far toward necessitating its expulsion from the force-imparting class of foods.

In further search for it, however, as an originator or conserver of force, the next point is to find whether it increases the excretion of carbonic acid (carbon dioxide) which, as usual, would take place chiefly through the lungs. This is a leading proof of the food-value of the hydro-carbons of oils or of such liquids as are changed so as to impart vital force.

This has been a subject of much investigation. Böcker, Perrin, Davis, and others have found a lessening of the amount exhaled (W., p. 105, and P., p. 273, note). Parkes considers it at least proven that there is no increase in the amount of carbonic acid. Edward Smith found the amount expired lessened by spirits as generally administered (see W., page 105, P., p. 273, and R., p. 72).

Dr. H. R. Wood, after a careful examination of authorities, says the " weight of evidence appears to be at present in favor of its diminishing the elimination of carbonic acid, although the matter can not be considered as entirely settled." More recent experiments still, and especially those of B. W. Richardson, seem to settle the matter. " As there is," says he, " a decrease of temperature from alcohol, so there is proportionately a decrease in the amount of the natural products of the combustion of the body. The quantity of carbonic acid exhaled by the breath is proportionably diminished with the decline of the animal heat. In the extreme stage of alcoholic insensibility, short of the actually dangerous, the amount of carbonic acid exhaled by an animal and given off into the chamber I constructed for the purpose of observation, was reduced to one-third below the natural standard " (p. 71). Similar experiments on man in the earlier stages of alcoholic derangement of function accord with this.

It in no wise bears this second great test of a heat-producing food. Nor does it in any way in the system show such affinity for oxygen as to form water.

So far from this it is a great water consumer in every tissue to which it finds its way. It dries the stomach, the liver, and the lungs. It even steals moisture from the very corpuscles of the blood, and so far interferes with the supply of water and with its value as the universal medium of exchange amid the tissues, that for this very reason it oftener than any other article in common use initiates degeneration of important organs (R., pp. 41, 42). With water and alcohol the endosmosis is toward the alcohol, and alcohol requires four times the pressure to pass through the same membrane that water does (see Dalton's Physiology).

When we consider how much all the functions of life depend on what the chemists and physiologists call osmosis, or the transfer of liquids with their soluble ingredients through various membranes, and that water is the all-important vehicle of this transfer, we must regard anything that interferes with this as involving serious risk both to function and to organ. A derangement of this all-pervading life-function is involved more in the continued administration of alcohol than in any one of the articles of the Materia Medica. It is fraught with imminent peril to the whole vitalized and vitalizing structure. So alcohol does not so combine with oxygen as to provide moisture and thus show one of the results of its appropriation by the system as an energy, but it makes grand larceny of the very thing it should contribute.

Alcohol thus far failing by any of the ordinary tests which apply to foods to vindicate its right to a place in their enumeration, another question frequently meets those who show its failure in these chief regards—to identify itself as an available food. *Is not Fat recognized as the evidence of a nutritive process, and does it not result from the use of alcohol?*

The answer to this is, first, that we lack full evidence to show that mere alcohol diluted with water has any power to produce accumulation of fat in or upon the other tissues.

When we wish to fatten, next to general sustentation we use those articles which contain fat, or those which are shown to be convertible thereto in the system. For instance, Dumas and Milne Edwards have shown that bees fed exclusively on sugar will still produce wax, which is a fat (see F., p. 825).

So, in the case of both starch and sugar, it is shown that in the human system these are convertible into fat. But alcohol neither contains any fatty material, nor have we been able to find any such conversion as is ascertainable in the case of foods.

Beers, ales, and some wines, when substituted for cheaper and better aliments which would serve the purpose, favor the deposition of fat just because they have the sugar, starch, etc., which can undergo this change. But to justify their use, it must be shown that they will accomplish this better than it can be done by any other method, and that the combination does in no wise imperil the most perfect nutrition of other tissue. The alcohol itself must appear as the indispensable factor. It is possible that the use of alcohol, by the tendency which it has to benumb the nervous system and to incline to *ennui*, may prevent waste by suspending exercise, just as geese may get fat livers by being shut up and fed in the dark.

But actual experiments show that under the use of pure spirits with the usual amount of exercise, neither man nor beast will increase in fat. Neither carbon or oxygen are separated therefrom in the method which happens to the hydro-carbons, and there is in it no fat to be deposited in the tissues (F., p. 825).

Again, if we could find a sufficient number of experiments with alcohol to assure that there had been increase of weight, in the absence of any evidence of *modus operandi* such as is relied upon as to fat-producing foods in general, we would need to be assured that the increase itself was not an adynamic result.

Parkes, in an experiment with another article, for instance, found that the subject " became indisposed and gained weight, as if excretion were interfered with " (see p. 256, note). A person may increase in weight by the growth of a fatty tumor, or a deposit of adipose tissue, by virtue of interference with normal and healthy functional activity of organs. We

2

would not be justified in calling such an article a nutrient.

Fat as an accumulation in the system, while often desirable and normal, is not so constant a sign of perfect nutrition as would be the development of muscle, or the steady increment of muscular power or vital force.

As we shall hereafter see, the most constant effect of alcohol seems to be to cause that fatty degeneration of organs which is a sad substitute for healthy alimentation.

If alcohol should ever be shown to cause increase of fat, with the facts of animal chemistry as thus far elucidated, it would be far more likely to be found to be a° pathological result of some obstruction to necessary change, than a healthy and vital contribution of force. We do know, indeed, that it does cause fatty substitution, or as Dickinson expresses it, " replaces more actively vital materials by oils and fibrous tissue " (Vol. LVI., Med. Chir. Trans., 1873). With the entire absence of fat material in alcohol, and of all those articles from which chemistry has found it to be produced, any increase would be due to such degenerative change, or, as Prof. N. S. Davis states it, " To a slow accumulation of hydro-carbonaceous material." (See, also, Carpenter). The individual fattened under such influences invariably diminishes in physical activity and power of endurance in proportion to the increase of weight." There is a fattening from abnormal " decomposition of tissues " (Dalton), as well as from normal processes. There may be a forced retention of the débris of the system which simulates the normal storing of vital force, but which is nevertheless so abnormal as to be damaging. We have known a man to grow fat amid the free daily use of opium, but this would not assert it a food. But the fact of science, uncontradicted by experience, is that alcohol by and in itself is not a promoter of corpulency. Such experiments as those of Hammond, in which the alcohol was taken with food, and an increase of bodily weight of less than one-third of a pound claimed, need only to be carefully read in order to show that a " proposition " is not " demonstrated " by such " series of investigations," because there were modifying contingencies not possible to be eliminated amid possible and probable sources of error.

Another claim that is instituted for alcohol as a food is, that when imbibed it does not fully reappear in the secretions, and, therefore, must undergo change in the system such as confers on it the nature of a food.

If alcohol passed away in the excretions so that every portion of it could be collected, this, indeed, would be almost demonstrative proof that it had served no purpose, either for nutrition or energy. But in the present state of quantitative analysis in animal chemistry, it can hardly be claimed that even if thus voided we would succeed in obtaining the exact quantity ingested.

It is not yet ascertained that we can collect in exact quantity any such substance taken, even though it may not serve any nutritive purpose.

On such a basis we would have to accept opium, arsenic, strychnine, and tobacco as foods, for these can not in quantity be recovered. Indeed, it almost seems strange that physicians should have urged this quantitative test with such earnestness.

It would be marvelous in any case that any such material passing though such a wonderful apparatus as that of the human organism, with all its methods of transformation and its manifold channels for excretion, should be recovered in exact quantity. We would have to test, in the most expert way, every part of the excretory apparatus for many days, and then even know how much still remained in the tissues.

Prof. Ford, of New Orleans, who is a careful experimenter (*New York Med. Journal*, 1872), speaks with the modesty of science rather than with the self-assertion of some clinicians, when he says : "Only exceedingly minute quantities of alcohol can exist in animal fluids and organs or be obtained from them, and, consequently, only the most accurate and careful processes and the manipulation of tolerably large amounts of material can enable us to procure that body, if it be present."

When as in articles like milk or eggs we find but a very small portion of the food material thereof in the excreta, when we know these foods to contain the very substances found in tissue, and are able to trace with accuracy the mode and degree of appropriation, it is quite proper to accept tables of food-values as approximations to scientific accuracy. But to accept as proof that such articles as alcohol, tobacco, etc., are foods, the single fact that we are not able to gather the whole amount ingested from the excretions is a strain for evidence which neither science nor reason can respect.

We find, too, that in some of the experiments relied upon, the kidneys alone were questioned and the skin not even examined (see foot-note, Parkes 271, Dupré). Some experiments with animals are necessarily only approximate. The skin, for instance, of a rabbit, with its fur, is quite different in its excretion from that of a man. A closed chamber like that of Pettenkoffer and Voit would be necessary to collect all the exhalation from the skin and the lungs, and the urine and fæces would need careful and repeated examinations, as well as the tissues. We look in vain for that degree of scientific accuracy which would have detected the whole amount even if rejected. Experiments show that the proportions excreted by the skin, lungs, and kidneys are not uniform, and that the time varies from twenty-four hours to five days. Richardson thinks that he has shown that the chief outlet is into the alimentary canal (p. 73), through decompositions and combinations which are not nutritive.

It has even been found by some observers that alcohol occurs in the secretions of those not using it as one of the débris of the destruction or decomposition of tissue (see Dupré, Béchamp [*London Lancet*, 1873,] *Lieben* in Parkes, p. 272, Ford, 1872).

Although the chromic acid test as modified by Dupré has much aided in the detection, it does not determine with accuracy the amount contained in each and all of the excretions. Anstie, in his last paper (Prac., 1874), says that it is not fully reliable (even qualitatively), because the chromic acid color test is affected by some substance found precisely as alcohol is (see also note of Parkes, p. 272). Indeed, from what we know of the laws of decomposition and fermentation, of the frequent results of some one of the alcohol series therefrom, several of them being produced from the fermentation of sugar (Fowne, 510), it is probable that alcohol is a slight product in the system much more frequently than imagined.

· The most significant fact that we can gather from the testimony as to the examination of tissue, organs, functions, and excreta, is that alcohol does appear unchanged in these to a much greater extent than would be anticipated from any considerable food or force-generator.

As is well known, Duroy and Masing (Parkes, p. 271) found such quantities of the alcohol in the secretions as to infer that all was excreted.

The fact, too, that it is found so largely in the blood and in the various organs after imbibition, proves that it does not undergo the usual form of digestion. An article found so plentifully in the tissues of organs, in the secretions, in the brain (Richardson), "in the spinal cord or substance of the nerves of animals which had been fed on it" (Hammond), is quite different from milk or starch or sugar or any of the chief nutrients. While there might then be an odor from it as from the most indigestible part of the onion, it would not be found in such amounts everywhere, and in the excreta if it were readily digested.

Other observers, and especially Anstie and Dupré, and Thudicum, failing to find in the urine such an amount as they concluded would be thus eliminated, infer that but a small portion escapes with the excretions. Their experiments do not seem to have extended to the other excretions. The experiments of Subbotin show that twice as much alcohol escapes by the skin and lungs as by the urine, and that this excretion continues much longer than most experimenters have supposed.

We do not presume that all alcohol imbibed could be detected in the secretions even if present unless all excreta should be examined and the time limit of retention amid tissues be known. We rest our belief that all the alcohol is not eliminated rather upon the unscientific unreasonableness of the expectation, than upon the completeness of the experiments had.

Such experiments to be conclusive would have to examine every excretion within a given time and know exactly what quantity should be taken in order not to interfere with physiological quantitative secretion, and that the person was in such a state of health that none of the alcohol could be lost sight of by reason of abnormal organism and function.

Yet strange as it may seem, the absence of this detection is attempted to be made the crucial test of the availability of alcohol as a food.

Not finding in it any of the elements of a nitrogenous food, finding none of those resultants which occur when starch, sugar, and fat are transformed into energy or vital force—unable to detect those secondary products which ought to occur if it is con- . sumed in any wise in the system, with the question whether it undergoes any chemical change yet unsolved (R., p. 66), it will not do at one leap-frog bound to assume that because all of it has never yet been collected from tissue and excretion, that therefore it must be a food.

Nor can we accept the statement : " If it does not escape from the body it must be oxidized in the body and must partake of the nature of a food " (Wood, p. 107). If so, the same must follow as to opium, quinine, and all toxics. It may undergo changes which are neither tissue-forming or energizing, and that which is not eliminated, unchanged, nevertheless escapes from the system as an excrement or inoperative for vital functions.

Prof. Parkes, of Netley, in his reply to Anstie, Practitioner, 1872, (p. 85), well expresses it thus : " Even if complete destruction within certain limits were quite clear, this fact alone would not guide us to the dietetic use of alcohol. We have first to trace the effect of the destruction and learn whether it is for good or for evil. You seem to think that the destruction must give rise to useful force, but I can not see that this is necessarily so."

When an article appears to a considerable extent in the excreta, it seems to indicate it as at least not readily assimilated.

When it could not all be found through experiments absolutely complete, instead of being very defective, we would next look for such results of disintegration as the laws of animal chemistry and chemical affinity would indicate ; and finding the changed products, would seek to know in what way it had availed as food. Finding none of those products, either primary or secondary, which attach to alcohol as consumed out of the system or such as occur when foods are appropriated in the system (see R., p. 64, and Ford, *N. Y. Medical Journal*, 1872, p. 574), we would be quite excusable for asking some evidence of food-value beyond the fact that it can not all be gathered as an excreta in form and substance.

We can find no chemical law or practical excretory result which defines it as a food. "The mode of destruction is, in fact, unknown" (Parkes, p. 272).

We search in vain for products of oxidation, except a slight change perhaps of a portion into acetic acid in the stomach, and in the case of the imbibition of large quantities the acidity of the urine was found slightly increased. "Yet in animals poisoned with alcohol, Buckhiem and Masing could find no acetic acid in the blood" (P., p. 272). No one has, as yet, been able to show a method of consumption such as ought to be manifested if this article acts as a food. If decomposed at all, "it is," says Richardson, "probably at the expense of the oxygen which ought to be applied for the natural heating of the body" (p. 72).

The fact that some advocates of complete elimination have not been sustained by such experimenters as Dupré, Anstie, etc., merely leaves undetermined a side issue, in nowise crucial as to the food-value of alcohol.

Next, those who would defend alcohol as a food and yet are unable to identify it as such by any of those tests which apply to the usual nutritive or energizing foods, would assign it a place among what are called *accessory* or *auxiliary* foods.

2*

It is first to be noticed that these terms are not intended to afford a convenient escapement for a promiscuous assemblage of articles assumed as foods, to which no other classification can be assigned. As used in a definite sense by its originators, it was intended to embrace articles whose action was fully recognized as an aid in conducting vital processes or as furnishing material for combinations, the dietetic value of which is well ascertained.

Water, for instance, is by some called an accessory food, because, as Carpenter expresses it, " It holds the organizable materials of the blood either in solution or suspension, and thus serves to convey them through the minute capillary pores into the substance of the solid tissues."

So " chloride of sodium holds fibrin and albumen in solution and aids in the absorption of fluids in the system and probably produces the chlorine of the hydrochloric acid of the gastric juice." Iron and various salts are found in the blood ; not as foreign substances, not as disturbing elements in health, but are also found as transferred to tissues or as existing in ascertained chemical affinities.

The action of fruits is quite as definite as that of albuminoids, for their fluid, their sugars, their acids are traceable. The citrates, malates, acetates, tartrates, etc., by their reappearance as carbonates, give that alkalinity to fluids which seems so promotive of imbibition and exhalation, and so of that osmotic circulation through membranes and tissues which preserves due equipoise of vital currents (compare Parkes.)

Indeed, modern chemistry is doing much to narrow

the class of accessory foods and to show that our inability to place many of these foods into either the
nitrogenous or hydro-carbonaceous classes, arose from
our want of knowledge more than from the incompleteness of the outline of classification.

For instance, condiments, so-called, are often mentioned as specimens of things demanded somehow by
the system, and yet not belonging to these classes.
Yet M. Papillon, in " Nature and Life," and " Odors
and Life," says that the chemistry of to-day reduces almost all the odorous principles either to hydro-carbons, aldehydes, or ethers.

If we turn to some such standard authority as Attfield's General Chemistry, we find that fruit flavors
and other essences and condiments are identical with
the amyls, and ethyls, and methyls (p. 389), and that
" Volatile and essential oils exist in various parts of
plants, probably as mere combinations of carbon and
hydrogen," and so as hydro-carbons easily available
for energy (p. 405), (Fowne, 492–500).

We find in coffee:—cellulose, 34 per cent.; fat, 10 to
13 per cent.; sugar and dextrine, and vegetable acid,
15.5 ; and lignine, 10 per cent.; and it arranges itself in the cinchona tribe. There is also a solid acid,
an aromatic oil, in small quantities, caffeine and ash,
the chief ingredients of which are potash and phos-
phoric acid (Parkes, p. 287).

Tea has most of the same constituents, and the
theine, or alkaloid, is found to be the same as the caffeine of coffee, which is a volatile alkaloid, and con
tains much nitrogen (see Johnson, etc.) " The products of the oxidation of caffeine, which have been

studied by Rochleder, are of considerable interest, inasmuch as both their composition and their properties establish a close connection between these products and the derivatives of uric acid " (F., p. 757).

Wöller goes still further as a result of his experiments, and says that the greater proportion of the nitrogen in tea, for instance, is not in the theine, but in a proteine like casein, and that it is a very nutritious food (Liebeg's Annaten, May, 1871,) (see Corwin, p. 281). While we know not fully the method of action, yet it is easier to see how these may awake nervous energy and furnish nitrogenous and respiratory food in a healthy way. We never find them to make those organic lesions which point to alcohol as essentially a toxic. Neither are these found circulating in the blood, nor the smell of tea or coffee pervading the brain as if the tissues were unable in any wise to appropriate it.

It will not do to make a place for alcohol among the so-called accessory foods unless it shows itself allied by close resemblances, and has in itself or in its ascertained products in the system, some vital probabilities. .

Not only does not alcohol establish a place as an accessory food according to any known laws of such foods, but it is very well known to inaugurate deterioration of tissue and to disturb the relation between the nitrogenous foods and water. " In appropriating the water so as to rob the tissues as already noted, it thickens membranes intended for that vital transmission of fluids known as dialysis " (R., p. 92).

This local " sclerosis, or a hardening of nerve-tissue, is one of its frequent manifestations" (see Hammond.)

III.

But the form in which the idea is now most promi-nently advanced that alcohol is, somehow, a food, is that it delays the metamorphosis of tissue, and, so, in a sec-ondary, but, nevertheless, effective way results in nutri-tion.

By the metamorphosis of tissue is meant that change which is constantly going on in the system which involves a constant disintegration of material ; a breaking up and voiding of that which is no longer aliment, making room for that new supply which is to sustain the life. Vital power itself is found to be a process of reparation and decay.

A usual division is to call the process by which food is converted into tissue *progressive metamorphosis*, and that by which tissue is converted into force *regressive metamorphosis*. For this latter the term destructive assimilation (Dalton, p. 325) is also used.

Both these processes are physiological, and the re-gressive metamorphosis or destructive assimilation is as healthful as the progressive process. It sounds conservative of health to say of a substance, that it delays the breaking down of tissue, but the histolo-gist or physiologist does not allow a substance which occasions such delay, to possess, because of that, either dietetic or remedial value. To increase weight by prolonged constipation is not a physiological pro-cess.

Speaking of this regressive metamorphosis or de-structive assimilation, Dalton says: "The importance of this process to the maintenance of life is readily shown by the injurious effects which follow upon its disturbance. If the discharge of the excrementitious substances be in any way impeded or suspended, these substances accumulate either in the blood or tissues, or both. In consequence of this retention and accumulation they become poisonous, and rapidly produce a derangement of the vital functions. Their influence is principally exerted upon the nervous system, through which they produce most frequent irritability, disturbance of the special senses, delirium, insensibility, coma, and finally, death."

The description seems almost intended for alcohol. To claim alcohol as a food because it delays the metamorphosis of tissue, is to claim that it in some way suspends the normal conduct of the laws of assimilation and nutrition, of waste and repair. A leading advocate of alcohol (Hammond) thus illustrates it: "Alcohol retards the destruction of the tissues. By this destruction, force is generated, muscles contract, thoughts are developed, organs secrete and excrete." In other words, alcohol interferes with all these. No wonder the author " is not clear " how it does this, and we are not clear how such delayed metamorphosis recuperates. To illustrate this action by the power of a magnet and then call this delay an " ultimate fact, which, for the present at least, must satisfy us," is, in plain English, to explain an unproven " delay of metamorphosis " by an inexplicable assumption. And this is science! Beale, in his recent book " On Life and Vital Action in Health and Disease," 1875, says: " Never was there a time like the present in which mere assertion has so excellent a chance of being accepted without real criticism. It is not, in my opinion, too much to say that the manner in which some statements, falsely regarded as scientific, are received and extolled, is almost a disgrace." If it could be shown that alcohol in any wise abates the cause of undue and pernicious metamorphosis of tissue, it would be quite a different thing.

It must be borne in mind that this metamorphosis is not a constant quantity. It adjusts itself to the amount of nutrition, to the amount of energy required to be put forth, and, even to a remarkable degree, is adapted to the reparative processes which are needed to overcome the disturbing actions of disease.

To take an agent which is not known to be in any sense an originator of vital force ; which is not known to have any of the usual power of foods, and use it on the double assumption that it delays metamorphosis of tissue, and that such delay is conservative of health, is to pass outside of the bounds of science into the land of remote possibilities, and confer the title of adjuster upon an agent whose agency is itself doubtful. What we know of usual interference with metamorphosis of tissue is just in the other direction than that of repair. For instance, " many of the anomalous affections classed as gouty or bilious disorders, are evidently connected with defects in the regressive metamorphosis " (P., p. 258).

If it is a delay in that change by which food becomes force, it is simply interference with the assimilative process. This, if needed, is better accomplished by diminishing the supply of food than by the use of a disputable toxic, which as yet has no well-defined function. If it delays the regressive metamorphosis or destructive assimilation, this is an interference so weighty and unphysiological that it would need for its vindication as a health-promoter the most substantial evidence.

Pappenheim, speaking of certain deleterious substances, says : " They prevent the regressive metamorphosis of tissues, and thus injure health " (P., p. 263).

Cameron, city analyst of Dublin, speaking of another article alleged to retard metamorphosis of tissue, says of it : " It is rather paradoxical to claim that an article retards tissue decomposition when it is chiefly recommended as a means of stimulating and increasing the action of the body " (see p. 279). " It seems," says another, " unlikely that alcohol would be applied differently during starvation and during usual feeding." To delay tissue metamorphosis in health, is best accomplished by avoiding that expenditure of force which calls it forth. If such delay is unavoidable by reason of the vagrant forces of disease or the too great demands of labor, to assume that, independent of curative agency in the one case, or of enforced repose in the other case, alcohol steps in and delays metamorphosis of tissue, is to assume for it a *rôle* which needs an indication of the method rather than an assertion of the hypothesis, to entitle it to respectful consideration.

Most advocates have not attempted this, but have satisfied themselves with the word " metamorphosis " —fit title for indescribable vagaries. We are compelled to address to alcohol the query in one of Dry-

den's plays, " What, my noble colonel? In metamor
phosis! On what occasion are you transformed?"

Hammond, one of the more recent of this school
of explanationists, we are glad to say responds to
the necessity of going further, and attempts an answer.

After defining progressive and regressive metamorphosis, and asserting alcohol as delaying destruction
of tissue or "regressive metamorphosis," he says:
"It is not at all improbable that alcohol itself fur-
nishes the force directly, by entering into combination
with the first products of tissue decay, whereby they
are again assimilated without being excreted as urea,
uric acid, etc. Many of these bodies are highly nitrogenous, and under certain circumstances *might* yield
their nitrogen to the construction of new tissue."
"Upon this hypothesis, and upon this alone, so far as
I can perceive, can be reconciled the facts that an increase of force and a diminution of the products of
the decay of tissue attend upon the ingestion of alcohol."

In other words, alcohol takes hold of excrementitious matter, resublimes it, extracts from it what has
escaped the usual processes of animal chemistry, and
empties this extract into the system, and so is food.
And all this is the mere suggestion of *modus operandi*,
and is called delayed metamorphosis of tissue, and
put forward as an explanation. Even as scientific
investigators, we need to watch lest, under such garbs
we cover ignorance instead of elucidating truth. Not
all the suggestions of experimentalists are a part of
their experiments, and some of their suggestions have

not about them even the sound of facts, unless it be
in the word factitious. Tyndall has ably written on
the " Scientific Uses of the Imagination." It has its
use in science because speculative inquiry may put us
on the track of reality ; but, unfortunately, there are
those who too soon welcome a thought into the domain of a proved fact, and forget the severe scrutiny
to which we have a right to subject experimental
research. The method assumed or suggested is totally
without evidence.

Having failed to identify alcohol as a nitrogenous
or non-nitrogenous food, not having found it amen-
able to any of the evidences by which the food-force
of aliments is generally measured, it will not do for
us to talk of benefit by delay of regressive metamorphosis unless such process is accompanied with something evidential of the fact—something scientifically
descriptive of its mode of accomplishment in the case
at hand, and unless it is shown to be practically desirable for alimentation.

There can be no doubt that alcohol does cause
defects in the processes of elimination which are natural to the healthy body and which even in disease
are often conservative of health. In the pent-in
evils which pathology so often shows occurrent in the
case of spirit-drinkers, in the vascular, fatty, and fibroid
degenerations which take place, in the accumulations of rheumatic and scrofulous tendencies, there is
evidence that alcohol acts as a disturbing element and
is very prone to initiate serious disturbances amid the
normal conduct both of organ and function.

To assert that this interference is conservative in

the midst of such a fearful accumulation of evidence as to result in quite the other direction, and that this kind of delay in tissue-change accumulates vital force, is as unscientific as it is paradoxical.

Dickinson, in his able exposé of the effects of alcohol (*Lancet*, Nov., 1872), confines himself to pathological facts. After recounting, with accuracy, the structural changes which it initiates, and the structural changes and consequent derangement and suspension of vital functions which it involves, he aptly terms it the "genius of degeneration."

With abundant provision of indisputable foods, select that liquid which has failed to command the general assent of experts that it is a food at all, and because it is claimed to diminish some of the excretions, call that a delay of metamorphosis of tissue conservative of health!!!

The ostrich may bury his head in the sand, but science will not close its eyes before such impalpable dust.

But there is some attempted proof. What is called delayed metamorphosis of tissue conservative of health, is claimed to be shown by a diminution of the usual excretions of the body. This is the evidence which those who suggest the hypothesis admit to be requisite in order to show that there is delay of regressive transformation. These usual excrementitious substances, as stated by Prof. Dalton, are—

        CARBONIC ACID,
        UREA,
        CREATINE,
        CREATININE,

Urate of Soda,
Urate of Potassa,
Urate of Ammonia.

The chief escapements are from the lungs and the kidneys.

While fæcal matter is partly made up of articles taken into the system which are not capable of any change, when we remember the relations of bile and of the various intestinal secretions, we find that it has many excrementitious substances. The skin, also, has its important functions of excretion.

It is also to be borne in mind that the various excretory organs are, to a degree, compensatory, so that under circumstances quite often difficult of identification and explanation, a material, ordinarily passed by the one, is largely voided by the other. This is pointedly the case in disease.

We are not able, from the amount in one secretion, to reckon with accuracy the probable totality. Observation has taught us that this variation has, in usual health, wide limits.

A thorough, scientific test, therefore, of the real amount of excretion can be satisfied with nothing less than the quantitative collection and determination of all the excreta for many days, in various persons and under definite circumstances. We are thus confronted with one of the most difficult problems in animal quantitative chemistry, and fail to find a sufficient number of experiments to found a theory.

This decrease of tissue change under the exhibition of alcohol is not, therefore, as yet, verified by those tests which the case requires.

When we look to the carbonic acid we find that it is not so far affected by alcohol as to lead to such uniformity of result upon which might be predicated the possibility of a nutritive delay in the metamorphosis of tissue.

Parkes, after summing up the evidence as to this point, and referring to those who have claimed a slight diminution of carbon oxide, says: " It can not be held as absolutely proved to lessen the excretion of carbon " (p. 276), (see also, H. R. Wood, E. Smith, etc.)

The fractions are so small that those who claim a slight diminution to have occurred need to show that it was due to alcohol, and was in no wise compensated for by other emunctories, or dependent upon disturbance of function, rather than a conservative adjustment:

To infer from the contradictory testimonies we have, that there is delayed metamorphosis of tissue; then, that this was needed, and that it is so promotive of health as to classify alcohol as a food, is a series of question-beggings too gratuitous for the science of facts to respect.

In each of these three regards there is defect of evidence. Incompleteness as to any one of them is enough to break the link.

The argument attempted to be derived from a diminution of urea, and the other excrements of urine, is equally incomplete. Attfield says (p. 43) : " It is impossible sharply to define excess or deficiency of the amount of urea in the urine."

The leading and reliable experts in these experiments are not in accord, even those claiming decrease of urea often notice it as very slight. Parkes says "The experiments of Count Wollowicz and myself prove that the metamorphosis of the nitrogenous tissues is in no way interfered with by dietetic doses." "The water of the urine and the acidity are slightly increased; other ingredients were found in my experiments to be unaffected, and I can not but doubt the statements to the contrary." Sydney Ringer, author of the Hand-book of Therapeutics, reckoned a slight diminution of excreta through the urine, but speaks of Parkes' experiments as so accurate that they must be accepted. He takes rank among the very first of physiological chemists.

Any one who will examine into the manifold (see Dalton's Physiology, etc.,) causes which may vary the urea, will see how indefinite must be conclusions drawn from the slight variation of the urea and other substances named. Still more, when it is doubted whether the substance is increased by disintegration of muscular tissue. Dalton says: "This is by no means certain, since, in a state of general bodily activity, it is not only the urea, but the excretions generally, carbonic acid, perspiration, etc., which are increased simultaneously" (p. 329).

The increase is not substantiated as occurring, and if it did, it would not prove that it resulted from a healthful delay of metamorphosis of tissue, produced by alcohol.

The generally accepted view that alcohol reduces bodily temperature, having deprived us of an evidence of food-value once relied upon, it is now claimed that this very reduction proves delay in the retrogressive metamorphosis of tissue, and so a food-value. But quinine has more power as an antipyretic or reducer of temperature than alcohol. Yet this does not show that it delays metamorphosis, or even that its medicinal action is by this mode. Prof. C. Bintz, of Bonn, says: "Alcohol is inferior to quinine as an antipyretic, since it induces dilatation of the capillaries and favors the formation of pus" (see also Conheim).

In order to show that this lowering of temperature is due to conservative delayed metamorphosis of tissue, it must be assumed as proved that it lowers the temperature independent of any action on the nervous system.

Because Bintz, in his experiments, found a lowering of temperature after cervical section of the spinal cord in animals, it does not seem to us to follow that the lowering must be entirely due to a delay in detruction of tissue.

The action of alcohol extends to every part of the nervous system and to all divisions thereof. Anstie regards its action as most decided on the great sympathetic or ganglionic system. One of its very first effects is as a paralyzer of the vaso-motor nerves, and even of those supplying the circulatory system in its remotest capillaries.

If there is a lowering of temperature when a formidable toxic is introduced, we are to study the laws of poisons and how variously they may disarrange that equilibrium of heat which is so marvellously maintained in the system in health.

We are not to assume that this is a result of metamorphosis delayed in a nutritious or conservative way. Plausible suggestions must not be asserted as facts of science.

Anstie ventures the thought that alcohol, "while itself oxidized in the blood with great rapidity, hinders the oxidation of the tissues" (see Practitioner, Vol. XI. 1873, p. 424).

That it is not readily oxidized in the blood is shown by the fact that so much of it is found unchanged in tissue and in secretions, and that neither the primary or secondary results of such oxidation are identified. In the absence of such results it more likely interferes with one of the most normal of all vital functions.

We should not have felt called upon to notice so fully this hypothesis that alcohol has accessory food-value by delaying regressive metamorphosis, were it not that this has been the explanation of its power most prominently and assiduously put forth of late.

All attempts having failed to identify it as a food according to any of those laws which attach to other foods, it is endowed with a kind of mystic or ethereal specialty of action, it being asserted that it delays "metamorphosis," and that, therefore, it is healthy.

It is fair that we have a full description of this "metamorphosis" which it delays. It is due that it be shown to delay it in accord with some well-known chemical affinity and how it does it. Next, that it does any such thing. Next, that if it did, it would be a nutrient, and not involve an accumulation of débris with real injury.

Ours is a day of philosophic phrases. When we

do not understand life, we go from tissue to cell, and from cell to molecule, and from molecule to atom.

Then, in search for plasm, it is easy to talk of protoplasm, as if the word explained the very point in question : viz., whence the productivity of matter ?

Just so vainly do we seek refuge from the nihilism of food facts as to alcohol by the cry of "metamorphosis."

When I was a boy of ten it was the first large word I had ever heard. Being given out to spell to a still smaller boy than myself, he spelled "metaporpoise." We have seen and heard it used since by those who, in its food connections with alcohol, get no closer approximation to its meaning.

No one can study this subject without being struck at the ingenuity of device—an ingenuity which is often honest, yet unconsciously resorted to to identify alcohol as a food.

The idea of its usefulness has been so far regarded as an accepted fact, that it seems to have constituted itself into an axiom which is to be verified by science rather than to be tested. But if, as Hodge expresses it, " It is unavoidable that the doctrines of theologians are largely determined by their antecedents and by the current philosophy of the day in which they live " (Vol. II., p. 397), it is not wonderful that the doctrines as to alcohol should be moulded by current antecedents.

The lamented Anstie, the devoted pupil of Todd, and schooled in all the enthusiasm of alcoholic fever-treatment, was led with Dupré to review the experiments of Lallemand, Duroy, Perrin, Massing, and

3

others who claimed that all alcohol was distributed in the tissues, or voided as excrementitious matter.

These experiments, we think, showed that former ones were inconclusive, and that the assumption that all alcohol passes unchanged through the system is not proven. But to conclude that it has undergone change operative for life-force, or has not in this change impeded some other vital processes, is quite an unauthorized deduction. In other words, the fact that chemists are not able to collect all that is imbibed being proved, we can not take the leap that when a material food or medicine taken into the system is not fully recoverable in quantity by chemical inquiry, therefore it must be a food. When it fails by physiological tests to show itself a food, we can not go out in hypothetical wanderings and assert that it is an auxiliary food in that it delays metamorphosis of tissue.

We can not do this without more definiteness both as to the meaning, the fact, and the desirableness of this delay. We can not accept ood-value on such a basis. And all the less because the diminution of urea, carbon, oxide, etc., claimed as evidence is not certified by most of the closest experiments, and neither the fact or evidence of the fact is complete.

Yet Anstie, Hammond, Johnson, Inman, and others seem to have adopted the hypothesis. Anstie, however, places the limit of the use of alcohol for the purpose of alimentation, so low as an ounce or an ounce and a half in twenty-four hours.

Johnson, in his "Chemistry of Common Life," speaking of ardent spirits of every variety, says ·

" They contain none of the common forms of nutritive matter which exist in our usual varieties of animal and vegetable food." Yet he copies the expressions of those who talk so much of delay in change of tissue, refers to " sugar and milk and spirit " as proverbially " old man's milk," and seems to accept the view that somehow it must be a food. We look, however, in vain for a single fact in chemistry which authorizes us to give it a place in " Common Life."

Most other authors seem to deem it sufficient to repeat this imaginary explanation, and merely *state* that the action of alcohol is that of checking tissue metamorphosis.

A recent Materia Medica makes of it a corollary in italics. Another asserts that " It offers itself in the place of the tissues to oxygen which would else feed upon them, and thereby retards their waste, while by the combustion of its elements with the same oxygen it becomes a source of heat." This unproved suggestion is offered as an actual explanation of the way in which alcohol might check tissue metamorphosis, and so doing be a food. Then the hypothesis is offered as evidence. Beale, who denies it a place as a food, and yet vindicates it as a medicine, says : " It may diminish waste by altering the consistence and chemical properties of fluids and solids " (see Powell).

Again he cautiously says that he regards alcohol as beneficial in some diseases, not by virtue of any really alimentary power, but because it arrests too rapid cell-changes (see, also, Anstie on Stimulants and Narcotics, 1865, p. 380).

In his work on " Protoplasm," p. 244, he says that

alcohol *probably* interferes with the multiplication of white corpuscles, and tends to prevent the disintegration of red blood-corpuscles.

It is important in quoting from scientists or from experimenters, to distinguish between what they claim as proved by their experiments, and merely hypothetical remarks or suggestions which they may make. A suggested effect must not be claimed as a result of research, when the author does not even put it forth as such. But as this is a suggestion of a possible method in which blood-change effects metamorphosis, let us inquire what there is in the action of alcohol that points us to a conservative power over the corpuscles of the blood which causes healthy modification of tissue change.

When taken into the stomach it acts as a precipitant to most organic compounds. Dundas Thompson says that "Alcohol added to the digestive fluid produces a white precipitate, and so suspends digestion" (see, also, Todd and Bowman's Physiological Anatomy, and Munroe).

Its action most certainly on the stomach is not that of a delay of metamorphosis, except of that progressive metamorphosis which takes place when food is changed into tissue. If used without large dilution, it certainly retards digestion, and when largely diluted is absorbed into the circulation as such. The microscope can trace an influence on the red and white corpuscles of the blood. It does, indeed, make commotion, changing the due relation of the corpuscles, often causing them to aggregate and depriving them of their usual functional power (see P., p. 45). The spherical is the form any mass of bioplasm will assume if it be free in fluid (" Life and Vital Action in Health and Disease," Beale, 1875). It interferes with this and with that passage of water through membranes so essential to the blood. Such interference is fairly presumed to be hurtful unless shown to be valuable.

If delay of metamorphosis of tissue means to disable the life forces of the system, in that sense alcohol may thus operate and be simply and positively injurious. E. Smith ventures the suggestion that all astringents may thus check metamorphosis by their corrugating effect.

It will not suffice to say (Stillé, p. 737) that alcohol does good by delaying metamorphosis of tissue, in that " it impedes the blood changes by which oxygen quickens nutrition, saturating the blood corpuscles and the liquor sanguinis."

The microscope says it impedes blood changes by interfering with the process by which the red corpuscles change off effete matter by means of life-giving oxygen ; that it dries the blood corpuscles by depriving them of the water needed for their life functions, and that while corrugating both the blood and the vessels, it relaxes them by its action on the vaso-motor nerves. These must not be summarily called nutritive delays.

Powell, Richardson, and others claim that by altering the blood globules it devitalizes them. Böcker, so often quoted, says that the blood of habitual alcohol drinkers, as yet in good health, shows a partial loss of power to become red by exposure to the air

in consequence of the loss of vitality in the portion
of the blood discs (see Powell, p. •151). Virchow
concurs with Böcker, and says that alcohol arrests
the development as well as hastens the decay of the
red corpuscles. We must not call this a delay of
metamorphosis of tissue, such as is conservative of
health, and exalt alcohol into a food. If this is
to delay metamorphosis of tissue in a conservative
way, then he nobly checks disease who initiates death.

Beale, as quoted by Anstie, says : " That alcohol
chemically arrests the vital cell-growth in nervous
tissue by coagulating its albumen, and in this way pre-
vents the too rapid waste of vital power " (Narcotics
and Stimulants, Anstie, p. 385). Also, " It incloses
the germinal matter in an impermeable coating which
no nutriment can penetrate from without " (p. 385).
Now, if alcohol can make such radical disturbance as
this, we want to know the exact therapeutic details, in
order to call such delay of tissue-action conservative
of health. Anstie himself expresses his dissent from
such a view. Yet by various authors we find such
checking of cell-change spoken of as if cumulative
of power for life. It is quoted and peddled from
book to book, as if it crystallized into fact by assert-
ive repetition, though still cloudy with hypotheses ;
and if true, an event evil and only evil, and that continually. We have studied " metamorphosis " and
cell-growth in the light of the researches of Billroth,
Wagner, etc., as occurrent in health and disease, and
we find nothing which confers on alcohol the agency
of a conservative delayer of tissue-change, such as
would classify it as a food.

We believe that any one who will candidly review the claims put forth for alcohol as a food, in that it delays in any of these hypothetical ways tissue-change, will conclude that it has no such power in any salutary sense, and that it is unwarrantably assumed that " to retard tissue metamorphosis is equivalent to tissue nutrition." We must say still more emphatically, in view of accumulated investigations since, what Dr. N. S. Davis well said, soon after these views were promulgated :

" It seems hardly possible that men of eminent attainments in the profession should so far forget one of the most fundamental and universally recognized laws of organic life as to promulgate the fallacy here stated. The fundamental law to which we allude is, that all vital phenomena are accompanied by, and dependent on, molecular or atomic changes ; and whatever retards these retards the phenomena of life ; whatever suspends these suspends life. Hence, to say that an agent which retards tissue metamorphosis is in any sense a food, is simply to pervert and misapply terms."

In a clinical study of alcohol we would far sooner claim that used well-diluted and in small and rapid doses and for restricted periods of time, it stimulates metamorphosis of tissue, and as an exhilarant sustains, while the system is thus supported during its attempted removal of pathological accumulations and its recovery of normal methods. Such a view needs proof, but facts are far more in that line of direction than in the other.

Another suggestion is, that alcohol may *belong to*

the catalytic substances, i. e., to those which by the " action of presence," as it is sometimes called, produce chemical change. Yeast, for instance, in changing sugar by contact into carbonic acid and alcohol, is said to exert a catalytic action.

In respect to this, it is sufficient to remark that no such power has ever been shown to attach to alcohol. The suggestion has only the appearance of science because it refers to a scientific fact, and with unscientific gratuity suggests an analogy unsupported by evidence.

" Pure diluted alcohol is not oxidized by exposure to air, but in presence of fermentative matter or of vegetable matter undergoing decay or change it is oxidized first to aldehyde, and then this rapidly absorbs oxygen and yields acetic acid " (see Attfield, p. 375). The aldehyde is a compound of which the formula is $C_2 H_4 O$, that of alcohol being $C_2 H_6 O$, and that of acetic acid $C_2 H_4 O_2$. The aldehyde would readily pass into acetic acid, and some experiments indicate that it so does in the system.

Here there seems to be an action of presence exerted upon alcohol, which, if proved to be salutary, might make of it a possible medicine, but would not elevate it into a food. As it is not thus discovered, it is more likely that the alcohol is thus changed into other products which are inert. So some of its evil effects may be overcome by its change into acetic acid, and its absence as alcohol in full measure in the secretions be all the more accountable. If the acetic acid is of any value, it could much more readily be furnished in vinegar.

Let us now in hasty review revert to the line of inquiry pursued as to the food-value of alcohol.

Our first search for its efficiency was among those
nitrogenous foods which chiefly contribute to the
welfare and production of tissue. With no nitrogen
of its own, and with no ascertainable action such as
is quite definite in the case of the usual nitrogenous
foods, it is not surprising that almost by common
consent it is dismissed from the plastic varieties of
food.

While there are well-ascertained laws which attach
to those foods which are prominently heat-producing
and force-propagating, alcohol can not be found to
act in accord with any method by which the human
system converts ingested articles into force. Animal
temperature is diminished rather than increased;
carbon oxide and watery pulmonary vapor are not
affected to any determinate and uniform extent. It
is even admitted by some who claim for it force, that
it does not arrange itself in the class of foods generat-
ing heat through oxidizing methods.

The interrogation point which Liebig put opposite
it as last in his list of respiratory foods is intensified
into well-authorized doubt as to its calorifacient
properties.

When we compare its chemical phenomena in the
system with those produced out of the body, we do
not get any evidence as in the case of foods (see
Fowne, p. 825).

While there is a class of auxiliary or unclassified
foods which are more and more being assigned to
well-defined classes as chemistry advances, we are not

3*

able to associate alcohol with these by any of the usual identifications. As the appellation is not intended as a refuge for all indefinable substances, we can not enroll it in this class.

Increase of fatty matter which is relied upon by some as proof of nutritive value, is not sustained as a fact where alcohol alone is fed, and is not in accord with experience as to animals. Even if for a time accumulation could be shown, it would not prove healthy nutrition to the same degree as increase of other tissue or of muscular force would, since alcohol favors fatty degeneration, and might possibly accumulate temporary fat amid mal-nutrition.

The hypothesis of delayed metamorphosis of tissue of a nutritive character is not well sustained either as to its possibility or as to its benefit if possible. Its effect on cell-growth is not so definitely ascertained as to be shown to be conservative of health.

Its catalytic action is not shown to be recognizable, as, for instance, is that of the presence of oxygen in fermentation (Fowne, p. 463), or of some nitrogenous body in vinous fermentation (p. 54), and is therefore purely hypothetical.

There have been some contradictory experiments, and some on both sides so inadequately conducted, as not to bear the test of examination or to commend themselves as veritable testimony.

There have been learned suggestions by scientific men which were not a part of their experiments, and which fail to crystallize about, or with, any ascertained facts.

The study of the laws of nature in the animal econ-

omy, and the results of chemical analysis, give us no
warrant by which we can certify alcohol as a food.

Dr. Markham, F.R.S., has well said : " It is scarcely
possible to read fairly the works of the distinguished
physiologists who have discussed the question, without feeling that they have been, spite of themselves,
as it were, driven by the legitimate consequences following from their premises to the conclusion that
alcohol is unnecessary and injurious to the human
body."

It is noticeable that this exclusion of alcohol does
not depend upon adherence to any rigid formula of
definition or artificial classification of foods. We have
for convenience referred to the usual divisions which
have been recognized as aids in determining food-
value and food-forces. Modern experimental science
seems to have rendered the action of foods even
more definite than was once supposed. Heat, for instance, as a mode of motion within and without, is
more fully recognized, and conservation of energy
becomes more explicit the more we investigate.

But in searching for alcohol as a food by all the tests
afforded to scientific investigation, it eludes the grasp.
Modern chemistry has substituted the crucible for the
alembic, and brought into animal chemistry all the
niceties of organic analysis. But still we must say,
" O, thou invisible spirit of wine." It may be the
fabled food of gods, but alcohol is not an actual food
for man which can be tried and proved such by any
known laws of any known science, or by any test of
any known art.

We are next brought to the question, whether al

cohol is shown to be a food by any of the tests of
dietetic and sanitary experience.

Experience is a noble word when it means anything. In an investigation such as this, it must be
borne in mind that it ought to mean something more
than the general run of a general impression. It
must mean the classified results of explicit and careful observation, in which sources of error have been
eliminated. Experience is really a series of experiments in which accurate observation plays a part akin
to accurate analysis. While nowadays it rightly
means that science with its experiments shall be
tested by art, it also intends that art with its experiments shall be tested by science. The experiment
of observation needs testing as well as the experiments of the natural sciences. They are congeners,
correlative of each other.

There is need of this caution because of the tendency of some observers, and especially of some clinicians, to scout at and dismiss scientific results not in accord with their experience, as if the latter were always
the sole arbiter of the other. They each sit in arbi-
tration, and should sit together. It is at their mutual
bar that truth and error await.

It does not do to prove alcohol a food by saying
that experience shows that all nations desire it.
This is denied (Parkes' Hygiene, p. 228), and if true,
would only show, as in the case of tobacco and other
habits, that man in his present estate has morbid
cravings.

The argument is like that of the optimist, who
makes of sin a virtue, because experience shows that

it is in universal demand, and therefore must be a form of good.

Nor will it do to cite cases in which it does no apparent harm. There are persons who live amid malaria and are unaffected. There are endurances of evils in exceptional cases which in nowise serve as experience. We call that experience which is the result of extended and classified evidence. So far from accepting individual cases, when the rule is shown physiologically, we seek the reasons of toleration, and often come to discover that the toleration itself is illusory, or at least affords no material for the enunciation or deduction of a law.

Nor will it do to speak of experience certifying alcohol to be a food when at the same time enough of real food has been used with it to sustain life. Cases (says Parkes, p. 201) are recorded in which persons have lived for long periods on almost nothing but wine and spirits. In most cases, however, some food has been taken, and sometimes more than was supposed, and in all cases there has been great quietude of mind and body. It seems very doubtful whether in any case nothing but alcohol had been taken ; and in fact, we may fairly demand more exact data before weight can be given to this statement.

Dr. Markham (in *Brit. Med. Journal*, June 14, 1862) shows that we must know not only the weight of the patient before and after the alcohol, but how much water, etc., was swallowed with it, and what rigid abstinence there was from everything.

Anstie, for instance, in endeavoring to substantiate his view of alcohol from experience, in his work on

" Narcotics and Stimulants," gives what he regards as evidence to this effect. From the ground taken by him it might well be supposed that we should have cases detailed with clinical accuracy. Instead of this, only one case under his own observation is given, and the others which have been mentioned to him are singularly incomplete as evidence. His own case he was not able to test, as the patient was not resident in the hospital. He came under his care with bronchitis, " half led, half carried." The daughter assured him that for many years he had lived on a little toast bread and gin, and he begged that " his drop of drink might not be taken away." A neighbor confirmed the statement. " The man's appearance was very singular and not easy to describe; it was not that he was very greatly emaciated, but he had a dried-up look which reminded one of that of opium-eaters " (p. 387). He smoked " a few pipes of tobacco each day," besides his " daily fragments of bread." Anstie adds in a foot note, " In my opinion the tobacco materially assisted to support life." Such is the case given to demonstrate that alcohol is a food. Bread, tobacco, gin. No detail of quantity, of proportions, or other exactness of clinical observation! Yet, amid the failure of any evidence such as attaches to foods, here is the evidence of food-power from its ablest advocate.

We find in Richardson this description of *gin*, which, even as to the liquor itself, shows how inexact such testimony as to alcohol must be:

" The spirit in common use that is most subject to the chemicals I have named is gin. Gin has to be made cordial, to be sweetened, to be rendered creamy and smooth, to be flavored, to be made biting to the palate, to be beaded, and what not else. To be made cordial, it must be charged with oil of juniper, with essence of angelica, with oil of bitter almonds, with oil of coriander, and with oil of caraway. To sweeten it, it must be treated with oil of vitriol, oil of almonds, oil of juniper, spirits of wine, and loaf sugar." After adding still further description, he quotes from " The New Mixing and Reducing Book " for dealers, as follows (p. 85):

" There is hardly any definite selling strength for gin, especially if it be sweetened. Within very wide limits no complaint is made by customers on the score of weakness, provided only the gin is creamy, palatable, and sharp-tasted. . . . . . Strong or unsweetened gin is in comparatively little request, and then with few exceptions only amongst the respectable or moneyed classes. At least three-sixths of the spirits sold over the counter of a public house consist of sweetened or made-up gin ; and as the sugar greatly alters the character of the liquor and deadens the original strength, it is possible for the retailer to consult his own interests by a liberal addition of water without in any degree exciting the disapprobation or injuring the health of those who patronize his establishment."

And Anstie's specimen took gin with bread, etc.

Such is the supporting evidence from *experience* that alcohol is so much of a food as that people may live upon it.

Dr. Slack also informed Dr. Inman " of two female

patients who loathed ordinary food and had subsisted
for months on nothing but alcohol *in one shape or
another.*" Another marked case became dropsical
and died at "the age of twenty-five," and another
lived " on ale, brandy, and water." With such looseness of statement and observation we are asked to
accept the food-value of alcohol. The author, on the
next page, even indulges in sharp reproof of those
who will not accept such evidence. We wonder not
at this when the author says in another place (p. 392)
that entire elimination of the alcohol " would not dis-
prove the possibility of alcohol acting as an aliment."

If the point claimed was simply that many drinkers
of alcoholic liquors consume less of the usual foods, it
would be granted at once. It is easy to see how those
who use beers, ales, manufactured spirits, wines, brandies, etc., may therefrom derive some both of nitrogenous and hydro-carbonaceous foods, how the ethers
may exhilarate, and the water serve as a menstruum
for real foods.

It is quite conceivable, with what we know of the
different powers of different persons to resist the effects of wrong food, drink, and surroundings, that some
will more or less imperfectly survive the use of alcoholics for a long period. Men live amidst miasm,
while others sicken and die, or indulge in intemperate
use of foods or drinks, and still attain age. There are
many who could survive a small dose of almost any
toxic for year after year, because the system somehow
adjusts itself to an evil, and establishes toleration, or
resists its full force. Because, as some claim, one person may eliminate a poison much more readily than

another, it will not do to claim that any such article is thus certified a food.

We have noted, for instance, that generally, spare men who commence drinking after twenty-five, are more resisting of the evil effects of liquors than those more robust, but it does not occur to us to attribute this to powers of alimentation in the alcohol when no such evidence is furnished. It is much more natural to assume that some survive, in spite of untoward circumstances, longer than others, rather than that the untoward circumstance itself perpetuates the life. This, too, amid admitted and appalling proofs of general death-bearing power.

The general law of direction of alcoholics is so pronounced and so adynamic, that science, not less than morals, has a right to put itself on the defensive, and to require close and positive evidence of exception where it is claimed. The practical test of real experience has forced it from its time-honored place in the ration of the soldier, until scarce a nation permits it, in the camp, on the march, or in the field. Because of the exposure on the Chickahominy in our late war, a half gill of whisky was allowed each soldier morning and evening.

Prof. F. H. Hamilton, in his "Military Surgery," says: "It is earnestly desired that no such experiment will ever be repeated in the armies of the United States. In our own mind the conviction is established by the experience and observation of a life that the regular routine employment of alcoholic stimulants by man in health, is never, under any circumstances, useful. We make no exceptions in favor of cold, or heat, or rain, nor indeed in favor of old drinkers, when we consider them as soldiers " (see " Military Surg.," pp. 70–75, etc.)

Insurance Companies, because of the facts afforded through their medical examiners, find the moderate use so deleterious as to give premium preference to the total abstainer, and exclude the occasional inebriate. A few cases of acquired tolerance or hypotheses of usefulness can not overset the physiological rule. While alcohol fails in the judgment of so many to vindicate itself as a food, and while animal chemistry utterly fails to indicate its use, although reliable in precision as to other foods, we must protest against attempts to clothe it with some ethereal efficiency to shelter it under indefinite experience, and to elevate it to the rank of a food so exquisite and subtle as to defy description.

We can not allow that persons who admit their utter inability to demonstrate its food-position, or to explain its assumed value, should, after a very few incomplete experiments, precipitate themselves into a climax.

" If it is not food, what is it ? "

We are able to establish other food-values, not with mere self-assertion, but with testimonies so veritable, so oft repeated, and so professionally exact, as by the very contrast to show the inadequacy of the evidence offered of the food-value of alcohol. *While failing here, it does not leave itself without a witness of its palpable pathological degradations.*

These evils are so pronounced that it is almost phenomenal to see how the advocates of alcohol as a food, as if to neutralize their theories, recount the direful evils which result from alcohol.

The *Tribune* Lecture of Dr. Hammond, as delivered before the New York Neurological Society, May 4, 1874, may fairly be taken as an expression of the views of the chief defenders of the food power of alcohol.

Commencing with criticism upon temperance reformers, " as those who have indulged in invective instead of argument, and whose facts are based mainly upon the immoderate use of the agents," he afterwards says it is impossible to say what immoderate use is; also, " That alcohol, even in large quantities, is beneficial to some persons, is a point in regard to which I have no doubt; but those persons are not in a normal condition, and when they are restored to health, their potations should cease," its use being " in the highest degree ruinous to society." He states that little value is to be attached to the fact that alcohol is detected in brain tissue, because he thinks the same might occur after a dinner of onions. Yet, afterwards, he makes much of its sclerotic action on the brain, and his original discovery of it in the spinal cord. " When I say that it of all other causes is most prolific in exciting derangements of the brain, the spinal cord, and the nerves, I make a statement which my own experience shows to be correct. I have already spoken of the remarkable affinity which alcohol has for the substance of which these organs are composed."

He then appends ten distinct diseases of the brain, and the eleventh encompassing " every variety of insanity, including general paralysis." Then follow three diseases of the spinal cord, four cerebro-spinal diseases, and five of the nerves, including neuralgia in all situations. He adds: "It will be noticed that sclerosis, or hardening, is a condition of all parts of the nervous system which alcohol probably often produces. It is doubtless the result of the direct action of alcohol on the nervous tissue."

Richardson, in his comments upon the sclerotic result, says: "The membranes enveloping the nervous substance undergo thickening, the blood-vessels are subjected to change of structure, by which their resistance and resiliency is impaired, and the true nervous matter is sometimes modified by softening or shrinking of its texture by degeneration of its cellular structure, or by interposition of fatty particles" (p. 110). "In addition to being the exciting cause of many diseases of the nervous system, alcohol probably predisposes to various others in which no direct relation can be traced. Neither does its action stop here, for the descendants of persons addicted to the excessive use of alcohol are liable to various disorders of the nervous system" (Hammond).

It will not do to get clear of all the testimony against alcohol by saying that this only applies to the "excessive use," and that the "abuse of a thing is no argument against its use." If there is no way of designating what constitutes excess, and if excess is the rule in those who attempt its beneficent use, surely as a question of public health and hygiene, as well as of morals, we have to do with a use which is so relative to abuse. The abuse of a thing is an argument against its use, if the very thing in question is, whether the article has any food use, and if the attempt to use it as such ends in abuse.

The vindication of alcohol as a food means more than its mere medical prescription. Physicians may be able to confine some medicines within the precincts of their own prescriptions, but we never have been, and never will be, able to place a food under purely medical control. Knowing the mighty injury to public health resulting from alcohol and from the conception that it has some food-value, we are called upon to place the real facts in this regard so far as ascertained, before the popular mind. It must not shield itself under the sanctity of natural demand, and so practically justify itself to be tolerated and legislated about as a food, unless amid the fearful and admitted risks it irrefragibly vindicates its place as such. Nor will it do, in order to defend " alcohol as a food," to claim that " none of us are living in a state of nature," that " hard work exhausts all the tissues of the body, and especially that of the nervous system ; " to more than intimate that the alcohol, which has such degenerative nerve affinity, is the thing to meet the case. Then call *dram*-drinking " a vile and pernicious practice ; " speak of " the inborn craving for stimulants and narcotics as one which no human power can subdue," and yet hope to confine alcohol to that kind of medical use which limits abuse ! If it is a food, we are willing to accept it in its totality, and while sipping wine at our own table, will not call him

" vile or pernicious " who seeks to *feed* upon his regular dram within the range of. his own ability (see Hammond).

Strangest of foods! most impalpable of aliments, defying all the research of animal chemistry, tasking all the ingenuity of experts in hypothetical explanations, registering its effects chiefly by functional disturbance and organic lesions, causing its very defenders as a food to stultify themselves when in fealty to facts they are compelled to disclose its destructions, and to find the only defense in that line of demarcation, more imaginary than the equator, more delusive than the mirage, between use and abuse. As a cause of disease, says Richardson, it gives origin to great populations of afflicted persons, many of whom suffer even to death, without suspecting from what they suffer, and unsuspected.

One of the leading defenders makes an ounce of alcohol in twenty-four hours the maximum. Another says, " A single glass of wine may be excess for some individuals " (Hammond), and that it is impossible to tell " what constitutes excess," except by personal experiment.

The very writers on the subject often seem to deceive themselves by their cautions, and to contradict themselves by their admissions. While informing us that " hypothesis by itself is the dreamiest of scientific rubbish," we can find proof in the same utterance that alcohol is only hypothetically a food. While warning us against excess, as the only security for life and health, they tell us that there is no possible measure of excess except in the trial. And the trial, alas, in the cases in which it is relied upon as a *food*, involves perils such as only the medical clinician may run, in those emergencies of sudden collapse or unconscious fever in which the peril is least.

We can not set persons thus at food experimentation for themselves upon an article which science and experience combined have not yet accredited as a nutrient, and whose medicinal value is still under adjudication. So long as the only indisputable facts are its potency as a toxic, and its physical, mental, and moral destructiveness as a beverage, we will not ask each man to test for himself what is excess, and so determine its food " use " or its food " abuse."

If a possible food in an accessory way under any possible circumstances (which, however, really means medicine), we might then start the question *whether it is ever a necessary food.* For since public health and hygiene have demands upon us as physicians, it would be legitimate to argue that it is untenable for us medically to sanction the use of a food fraught with so much peril unless it is a necessary food. When it is registered by its defenders as the most doubtful of all foods ; when admitted to have no determinate food-value ; when the cases of its asserted indispensability plainly limits it to the rôle of a medicine ; when the tendency to excess is admitted as the most prevalent, the most appalling, the most irresistible proclivity of civilized humanity ; when in all its sanitary and hygienic bearings it is by common consent registered as the greatest of all burdens on the public health, we might well pause and ask

whether it is a necessary food. That which can not
be proved to be a food at all might naturally be assumed to be unnecessary. When, in addition, you add
that to most of those using it as a food it becomes
the chalice of disease and a fearful impeachment
to health and life, the word unnecessary becomes intensified into injurious. Practically, the theoretical
food becomes labeled poison. He is the true physician who deals with it under the trade-mark extra-
hazardous, rather than he who vaunts it as having
somewhere, somehow, sometime, some ethereal attribute like the ambrosia which made Tantalus immortal
only that it might make him miserable.

It is well that such authors as Dupré, Anstie, Johnson, Hammond, and others, in fealty to experience, add
so many facts as to the evils of abuse. The utter
failure to identify it as a food at all from the stand-
point of chemical and histological research is enough
to dismiss it from the sphere of scientific support.
Vain is it to account for popular fallacy and former
medical opinion by saying it may check cell growth,
it may constrict lax tissue, it may delay metamorphosis, or may at least resublime the effete disintegration
and extract from it new life. Such hypotheses have
not even the aroma of scientific fact, save what is
given to them by the respectable names of those who
venture such *suggestions*, too often distorted into
scientific results by too zealous seekers after supporting authorities. We hold that it is a duty we as
physicians owe to our profession, our patients, and to
the interests of public health that we rightly estimate
and help the people to estimate the facts as to the

food-claim once put forth in defense of alcohol. We shall, in another section of this paper, attempt to weigh its value as a medicine for professional prescription. But if the laity are determined to use it as an assumed food, or as a self-prescribed medicine, let them not quote us as authority for either.

The relation of voluntary associations or regulative or prohibitory legislation thereto, it is not our special province to indicate.

But to know and state what science and art prove or fail to prove as to the food-value of alcohol is our province. To expose fallacious arguments for its use and to point out the physical entailments involved is also our proper function. A world-wide constituency is entitled to know from the chemist, the physiologist, and the practitioner whether there is any food-demand for an article the use of which beyond dispute is fraught with the most prevalent and direful results to the physical structure.

When the consumption in this country (see Dr. Hargreaves) amounted in 1870 to 72,425,353 gallons of domestic spirits, 188,527,120 gallons of fermented liquors, 1,441,747 gallons of imported spirits, 9,088,894 gallons of wines, 34,239 gallons of spirituous compounds, and 1,012,754 gallons of ale, beer, etc., or a total of 272,530,107 gallons for 1870, with a total increase of 30,000,000 gallons in 1871, and of 35,000,000 gallons in addition in 1872, it certainly is a great medical question as to what influence this immense consumption is exerting upon the public health. Such facts inherently inquire of us what impression these alcoholic solutions are making upon the body

4

for sustentation, or how far they operate as the causes or abettors of disease or death. (See, also, Hitch-cock, " Public Health," Vol. II.)

In one sense we seem to anticipate the sphere of the political economist, since what is expert and profes-sional is fundamental to what is more general.   In a day when science means facts carefully collated, and when experience means not general opinion, but re-sults proved by tabulated clinical evidence, we must seek to reply with no uncertain sound. It is apparent that it is too often Bacchus that would appear in the robe of Ceres, or with the wand of Esculapius, and so call that food or medicine which is taken merely as a luxury or to satisfy an appetite.   We shall have done essential service when in our own peculiar sphere we shall have removed all the false supports for alcoholic beverages, attempted to justify its use as a daily or occasional self-selected food or self-advised medicine. We ask no concealment of facts, no distortion of evi-dence, no yielding to preconceived sentiments, since the lines are being drawn so closely as greatly to . narrow the indications for alcohol in any department open to our advisement.

If still it must initiate disease, confirm ill-health, shorten life, and accomplish death, it can not with medical sanction be palmed into respectability under the guise of a necessary or defensible food.

PART II.

IS ALCOHOL A MEDICINE?

In this second part of our subject we propose to discuss *The place of alcohol in the Materia Medica and its medicinal action.* "There is," says Lionel Beale, "no more important question in medicine to be determined than the action of alcohol in disease."

The recent address on Practical Medicine by Dr. Austin Flint (1875) before the Am. Medical Association, recognizes the saliency of the inquiry, as does the common sentiment of most practitioners.

As physicians, we can not ignore the fact that it is the article of the Materia Medica direst of all when it escapes the bounds of medical necessity.

It is the medicine which is most prone to overleap all barriers, and so often glides into the sphere of lustful appetite that it numbers victims by thousands within the pale of ebriety, and by tens of thousands beyond it.

If I knew that brandy would save my patient, and that a thousand copying from his restoration would make self-resort to the same remedy and die, I would in solemn sorrow, yet in holy fealty to my patient, give him the brandy, and hold myself not responsible for the self-inflicted result to others.

But when I come to know that the remedy itself is under trial as a remedy at all, that equally efficient substitutes are claimed, that the so-called self-infliction is so infatuating that it proves a swift delusion to many of the wisest and the best, both my profession and my manhood require me to bring it to the most rigid tests of necessity. If I can not ignore it, I must at least set bounds to its vagaries. I must use it as a medicine so as to secure the greatest good to the greatest number so far as is consistent with the interests of the patient in hand.

As we come to inquire into the value of alcohol as a medicine, after having found it unsustained as food, it is well to remember that the terms, food and medicine, are often more nearly allied than the mention of the words is apt to indicate. A medicine is that which helps to heal or repair, for that is both the etymology of the word and the practical design of the article used. The process of restoration or repair is often but an application of the process of natural nutrition. Amid the progressive change of food into tissue, and the regressive disintegration which all life means, we must not deceive ourselves by terms. Much of the discussion, therefore, as to the value of alcohol as a material for medicine, is in reality to be determined by what it can do toward repairing the waste of tissue which occurs in disease. What it can accomplish in this regard is largely the determination of the question of its food-value. This we have seen to be so small and indeterminate that *it will not do to push it forward* as a valuable medicine in those regards in which a medicine chiefly concerns nutrition and the production of animal heat. " The more we investigate," says Lankester, " the relations of food to the human system, the greater must be the conviction that food is not only capable of maintaining healthy life, but by proper modification can be made the means of curing disease;" and again, " In the management of food we have the great means for the cure and removal of disease."

When we find that alcohol has no nitrogen with which to nourish; that it does not respond to the laws by which animal heat is usually evolved; that it at best undergoes such imperfect change in the system that much of it is found unchanged in secretions, excretions, and tissues; that the products of its primary or secondary change can not be identified; that it is not settled whether it diminishes the carbonic acid, or urea, or other excreta, or that such diminution would be reparative, we may well hesitate to assign it a place in the category of medicinal nutrients. While it has eluded the ingenuities of science, the persuasions of art, and the astute diligence of interest to extemporize it into a food, it has failed not less signally to vindicate itself as a medicine in the one particular in which it is most frequently urged as of value.

Another consideration that should make us exceedingly modest as to clinical assertion of its value as a medicine is, that the very next important thing to nutrition of tissue which it has been claimed to accomplish, viz., the sustaining of animal heat, is now the very thing which it is claimed to reduce. We have examined with some interest the different records of the value of alcohol furnished by authors on Materia Medica and Practice of thirty years ago and those more recently in print. New editions of the same author show often an entire shifting of the line of defense, or else repeat the old dicta in ignoring ignorance of the most defined views of modern analysis and experiential use. Many others have come to deny its value except within very narrow limits.

From being the vaunted remedy which experience had relied on for sustaining animal heat, if used at all it is to lower it. Even tea and coffee have taken its place as general stimulants. Caffeine has a far more definable action, as overcoming fatigue and sustaining the body during undue exertion, than has alcohol. The experiments of M. Robuteau go far to show that they are not, like alcohol, mere temporary stimulants, but contribute directly to tissue and force (see Cameron, p. 281).

In the face of some of the most precise experiments made by those who view this article from a purely scientific stand-point, and in face of the experience of those who have clinically used the thermometer and the sphygmograph, and satisfied the profession at large as to its non-increase of temperature, we can scarcely understand how the new edition of a Materia Medica published years since should say, "that no one has successfully controverted the proposition that the temporary stimulus enables men to exert muscular energies far beyond their habitual power, and that more than anything else it repairs the exhaustion of fatigue and reanimates the body about to perish from cold."

It will not do to assume unanimity of testimony, and so carry a position, when page after page can be quoted from medical authors in nowise identified with any special reform, in opposition to such views (see summary by Parkes, pp. 280–285).

An examination of the leading books on Materia Medica and Therapeutics within the last twenty years, and of the limits which both science and practice have demonstrated, will serve to show how narrow is the field in which we are to look for the therapeutical effects of alcohol.

Excluded by common consent from the list of ordinary aliments, eliminated from most of modern dietaries where foods are studied with precise relation to force and effective endurance, and from all systems of athletic training, pronounced so unreliable as a sustainer of animal heat as to be used on a directly opposite hypothesis, identifying itself with toxics to a degree that almost organizes use into abuse, its field is so far narrowed as that the only classification it will admit is that of GENERAL STIMULANT. Those who still speak of iodine or mercury as an *alterative*, now find no such place for alcohol. It is thought to have a secondary sedative effect, yet with this are associated such other results, that it is seldom prescribed with that view.

The narcotism which follows its stimulation is rather a paralysis of vaso-motor and general nervous power, and will not bear favorable comparison with those real sedatives and narcotics whose therapeutic agency is now so much more generally invoked where such an effect is desired. That it does temporarily increase

heart-action when given in moderate doses is generally admitted. Even this, however, seems to have
very narrow limits. Richardson seems to have clearly
shown that its chief influence as a stimulant is that by
its action on the " organic nervous chain," it produces
paralysis of those nerves which supply the blood-
vessels. So the vessels being relaxed, the heart
beats with " increased frequency, with a weakened
recoil stroke " (p. 50).

The experiments of Parkes not only confirm this
view, but show also how soon the heart loses power
and becomes enfeebled by this disturbed action. He
says : " In the exhaustion following great fatigue, alcohol may be useful or hurtful, according to circumstances. If exertion must be resumed at the expense
of the heart's nutrition, then it may be used, but only
in small amounts, and with Liebig's meat extract "
(p. 280). When renewed exertion is not necessary, he
advises to let rest recruit the body, and to avoid hurtful quickening of the heart by alcohol.

The editor of the *British Medical Journal*, 1874, p.
457, in an able article on the researches of Parkes,
says : " Many a horse which might have reached its
journey's end, at a snail's pace, it is true, but still
safely, has utterly broken down on the road in consequence of a too frequent application of the spur."

Yet in some sudden attacks of faintness, as resulting
from failure of heart-action, or some profound nervous
impression conveyed to the heart, alcohol may cause
a reaction, and if no other article is at hand, may be
never so good as a medicine in such an emergency.
Yet an excellent recent authority says that it should

not then be relied on alone, but "with it should also be exhibited some more rapidly-acting, diffusible stimulant, such as ammonia" (p. 107, Wood). For, it must be borne in mind, there are attendant evils. The blood discs are disturbed, their vitality diminished by interference with their water, and their interchange of oxygen and carbon-oxide embarrassed. The agglutinated cells do not pass so readily to the extreme capillaries (see Richardson).

The blood, as Beale thinks, is rendered less fluid, and the vessels corrugated (see Protoplasm, 1870). If this is an advantage, as he thinks (not so Anstie), it can not be in usual faintness unless occurring from the loss of blood. Poiseulle, on the other hand, has shown (see Letheby) that it is a physical as well as a chemical and physiological agent, for it hinders the flow of liquids in narrow tubes, and may act in the same way on the movements of the blood in the capillary vessels.

Powell and Böcker show that the blood becomes poorer in iron, and that the blood discs are devitalized from its use.

Richardson, in his recent work on "Diseases of Modern Life," says: "The heart beats faster because the contractile force of extreme vessels is weakened, and so there is less resistance than natural." "It gives evidence not of increased, but wasted power" (p. 218) or as Parkes puts it (p. 273), heart nutrition is "interfered with" (so R., p. 90).

Men and animals killed after imbibing largely of alcohol, show a most congested condition of lungs and brain, and other organs are involved. This re-

4*

laxation of arterioles and capillaries is a double-headed power, and needs the strictest limits as to the time it is to be kept up. The records of alcoholism, and the milder external records of moderate drinkers, show unmistakable lesions as a result of such engorgements, with subsequent change of structure and impairment of function. Even the faces of those who are moderate drinkers often show its influence on the capillary circulation, and how it deranges the normal movement of the blood through the blood-paths of the system.

The limit, too, in quantity is so pronounced that $1\frac{1}{2}$ ounces in twenty-four hours is admitted by such men as Parkes and Anstie, to be the maximum for use where any lengthened continuance is needed (see Parkes, p. 277).

Concessions which pathology has, by its facts, made necessary, may well lead us carefully to scan its effects in the most reduced quantities.

The effect upon the circulation, which is the most prominent and declarative one, is the very one which is so often shown to be disastrous.

Those who have studied most carefully the history of drinkers, find that local congestion is really the most usual result of the steady moderate use of ardent spirits.

An article which paralyzes the vaso-motor nerves, which causes the capillaries to receive more than their usual amount of blood, to receive it with its flow complicated by changes made in the fluid itself, and which still shows alcohol carried as a foreign, toxic substance through the circulation, may well demand a demonstration as positive of its good effects in some cases as it gives of evil effects in others. Even Anstie, in one of his latest papers, while claiming a " theoretic basis for the use of alcohol (could we find the right form and dose) in the early stage of acute inflammations, with a view to prevent the migration or the too rapid destruction of blood-cells," adds: " Whether in practice the effect which we desire can be really attained, is a point that I do not think has been quite settled as yet " (Practitioner, 1873, pp. 361–8).

Besides the effect of alcohol as a stimulant to the circulatory system, some claim for it a value as a nerve stimulant.

Dr. Hammond can not be supposed to give us an overdrawn summary of its office here. The fearful catalogue of ills and the marked tendency to pass unchanged through organ and tissue, to induce sclerosis and other change of structure and to modify normal circulation and exchanges, as before noted, is so weighty and critical, that the word abuse does not settle the demarcation as a medicine any better than it does as a food. There may be great abuse of an article as a medicine, and because so used it does not necessarily suspend its destructive agency. Where there is no separating line, the very use has in it so much of the spirit and power of abuse, that it needs the eye of special insight and the tactile judgment of satisfactory experience to vindicate the risk. It is " of all other causes most prolific in exciting derangements of the brain, the spinal cord, and the nerves " (Hammond). " The effect had on the nervous centers starts them directly in the° path of nervous exhaustion (Richardson). As its direct action is " to lessen nervous force (Edward Smith), and as it is an irritant of nervous tissue, it is difficult to dissociate its vaunted nerve-stimulation from nerve-irritation." Anstie claims its chief effect to be on the ganglionic or sympathetic system, while J. C. Morris (*Phil. Med. Times*, 1875) says that it acts directly on the cerebro-spinal nervous system, although afterward he attributes part of the effect to associate ethers. A most able chemical experimenter recently says, " If it is beneficial to man, it must be chiefly due to its effect on the nerves ; if it is detrimental, this is due to an effect of the same sort." Its action as a neurotic is its greatly predominating result " (Letter H. B. P., 1876). We believe this whole question as to the medical value of alcohol is fast reducing itself to the question of its direct effect upon the nerve centers.

Amid these, it makes some of its most direful records of toxic effects. We are scrupulously to search whether the effect can be so modified by dose, or is so altered in disease as to make that which is an enemy in peace, in war a friend.

With the stern catalogue of ills from the hands of neuro-pathologists before us, it behooves us to put alcohol on the defensive to establish its claims as a curative neurotic. *The next prominent defense of alcohol as a medicine " is to aid in the assimilation of food"* (Wood, p. 110). The phrase has been so often used, and the general popular and professional impression like that of its heat power has been so prevalent, that men cling to this view with the tenacity of an anti-

quated error. Pathology appears and tells us that
alcohol coagulates albumen, and that it acts upon
digestive fluids in a disturbing rather than in a beneficial way.

In very small quantities, and in a very limited degree, it may serve to castigate and so arouse a torpid
stomach, but facts as to its corrugation of tissue
(Beale), its tendency to excite mucous catarrh (Wilson Fox), its appropriation of moisture and its causa-
tion of enfeebled action in all glands (see Fox and
Parkes, p. 273), show how very restricted is the limit
in this regard. It is often used at the expense of
organic validity and causes chronic impairment of
function. We can not use it as a digester as we do
pepsin, and so attest its agency, but can only claim it
as an appetizer, as we do a condiment which caters to
a morbid or suspended appetite. Indeed, we shall
hereafter see that much of its credit in this regard
depends upon the ethers or essential oils so often administered with it, while the tonicity imparted depends upon nutrients taken at the same time.
Science, at least, has failed to indicate how alcohol
in any one way aids assimilation. Just so has exact
practice failed, and testimony rests too much on a
general impression and clinical belief. This, though
worthy of respectful consideration, recent investigation shows to be suspicious for want of accuracy of
evidence. We need clinically tabulated and classified
results from which sources of error are eliminated on
which to base any extended use of alcoholics for medicines. Such expressions as that it " bridges over
weakness," " assists assimilation," " acts as a quick

nutrient," etc., can by no means pass as axiomatic in the face of the fact that they do not stand the test of dietary studies, and are not accepted by many of the most advanced professional thinkers and clinicians of our day.

It is marvelous that a recent author should scatter in among his grounds of approval various deliverances chiefly supported by positiveness of statement as if they were accepted formulas and then cap the climax with the statement, that " The addiction to alcoholic drinks which is universal could not exist without some good reason." If this is to vindicate it as a material for medicine, surely the *argumentum ad hominem* is equivalent to the *reductio ad absurdum.*

Prof. H. R. Wood, Jr., presents, in compact form, the chief claims made by others for alcohol as a medicine.

" In the advanced stages of disease, when the typhoid state is well-developed, then alcohol should be given boldly, to quiet by stimulation the nervous and circulatory system, to afford a food which shall in a measure replace the natural pabulum, to aid in the digestion of milk and other simple nourishment, to aid in lowering temperature by checking the tissue-waste of fever ; in a word, to enable the system to stand the drain upon the vital powers, and at the same time to check such drain."

But as we analyze, what do we find? The very therapeutical effects which it is sought to secure therefrom still under question as at all possessed by this drug.

When we seek to quiet by stimulation the nervous

and circulatory system, we are using an agent which is only anæsthetic in excessive doses and paralyzing rather than sedative. "Under such conditions there is an interference with the ordinary nutritive processes" (R., p. 61). From what we know through later pathology of its effect on the circulation and on nerve-centers, it is not the best article for such sedation.

Another office assigned is "To afford a food which shall in a measure replace the natural pabulum.' Whether it does so at all is the very thing in question. If milk and simple nourishment are not digested without it, we have no therapeutical or alimentary explanation of how it can avail to replace or aid them.

Prof. Voit, who does not ignore it as a medicine, and whose experiments as to it have been among the most exact, says in substance, "that alcohol, when freely given, causes such disturbances that we can not give it in amount sufficient to be considered a nutriment" (see Alcohol as a Nutritive Agent, *Boston Medical and Surgical Journal*, June, 1872, IX., p. 413).

As to "lowering temperature by checking the tissue-waste of fever" it is too rapid a generalization to assume that a lower temperature always betokens a healthy arrest of tissue-waste; that it is the alcohol and not the associate ethers and liquids and other remedies that in part accomplish it, and that a better thing for this purpose can not be used.

The ill-conduct of alcohol as a general febrifuge at once limits such an argument. Notice, too, that it thus comes to be claimed as at once the stimulant which hastens progressive metamorphosis, and the sedative which delays regressive metamorphosis, the method and the fact being alike unknown and the results predicated being contradictory.

While we realize in alcohol the one capability of acting as a stimulant where there is sudden failure of heart-action, or where by reason of impeded circulation the blood fails to reach the capillaries in due amount, this sphere of use is exceedingly limited both as to time and amount.

In all cases where day after day this effect is sought, we are in danger of some of the accompanying ills of alcohol, and find ourselves more and more able to rely on water and ice and quinine, as antipyretics; on milk, eggs, and other nutrients; on ammonia, essential oils, and ethers as exhilarants, and on emulsions, cod liver oil, and other fats for a quick replacement of wasted tissue.

Upon recumbent rest as an aid to diminished force in the heart and blood tubes, upon the bromides as quieting the nervous system, upon sedatives or narcotics such as induce sleep and make of it the real recuperant, and upon that aid from food which resists all usual adynamic tendencies and compensates for tissue-waste, are we to rely, more than upon a cardiac stimulant or a possible nerve-exhilarant or a digestive excitant.

[Perhaps typhoid fever is the disease in which the profession in general are most disposed to avail themselves of the use of alcohol. Since the meeting of the International Medical Congress, Prof. A. L. Loomis, of New York city, has published several of his lectures on its pathology and treatment.   No one in our country can speak more authoritatively, and as he has no radical views as to the exclusion of alcohol, it is worth while to notice the place to which he assigns it.   In the milder cases, he entirely excludes it. As a means of reducing temperature, he does not mention it, but relies on cold, quinine, and sometimes digitalis with quinine.   He says, " There are other antipyretic agents besides these two, but they are of so little importance that it is necessary to give them only a passing notice."   He then states that, about the third week, signs of failure of heart-power may begin to manifest themselves; the pulse become feeble and irregular, the surface cool and moist, the patient exhausted and unable to turn in bed, and the tongue have a dry and brown appearance.   Even when there is need to sustain heart-power and to use some form of stimulants, he gives the following rules :

"*First*.   They should never be administered indiscriminately—that is, never give a patient stimulants simply because he has typhoid fever.

" *Second*.   Where there is reasonable doubt as to the propriety of giving or withholding stimulants, it is safer to withhold them, at least until the signs which indicate their use become more marked.

" *Third*.   In every case, but especially when stimulants are not clearly indicated, watch carefully the effect of the first few doses.   There are few whose experience in the treatment of typhoid fever is such as to enable them to positively determine, from the appearance of the patient, when the administration of stimulants should be commenced."

He then insists that the patient should be seen every two hours to watch the effect, and unless the pulse becomes fuller and more regular, the first sound of the heart more distinct, the restlessness and delirium less, the tongue more moist, and the patient more intelligent, the remedy is contraindicated. While highly appreciating a possible value, he thus shows how deleterious is a less guarded use. While we approve of such employment, yet it is not to be concealed that many excellent practitioners rely wholly on ammonia, ethers, and foods in such cases. Loomis then takes occasion to show the entire reliance which must be placed on milk and other nutrients to repair the waste of tissue.]

In all Chronic Diseases, the place of alcohol, as a remedy, is, by quite general professional consent, still more restricted.

The poor inebriate no longer needs to be assured that a hair of the dog that bit him can avail to cure the bite. Few would now say with Anstie that alcohol is " good to prevent epilepsy ;" that it is " the best treatment for the convulsions of teething ;" the sovereign remedy in neuralgia, the cure-all for dyspepsia, and the eradicator of tubercle in phthisis.

Facts of pathology, as to brain, and nerves, and stomach, and lungs, have of late come in upon us too rapidly for this. Fatty and fibroid degenerations in every organ of secretion, warn us as to the serious import of the embarrassments which it is the normal tendency of alcohol to initiate and confirm. Richardson, from his experiments, attributes the deterioration of tissue, to a disturbance of the relation existing between the plastic colloid of nitrogenous foods and its combining water.

" These alcoholic degenerations," said the army professor at Netley (p. 276), " are certainly not confined to the notoriously intemperate. I have seen them in women accustomed to take wine in quantities not excessive, and who would have been shocked at the imputation that they were taking too much, although in their case the result proved that for them it was excess." This question of excess occurs in sickness as well as in health, and all the more because its determination is so difficult and the evil effects so indisputable. The dividing line in medicine, even between use and abuse, is so zigzag and invisible that common mortals, in groping for it, generally stumble beyond it, and the delicate perception of medical art too often fails in the recognition.

Yet those evils which impress us so much as physicians that we can scarcely reach the questions of morals, are brushed decorously aside with the assertion of innocence and benefit, if kept within proper bounds. This is but a jugglery of words where our science and our art are so uncertain as to the boundary. It substitutes the dicta of conscientious empiricism for such guide and evidence as we need where such weighty interests are pending.

The therapeutic effect of remedies is nowadays being studied with such precision as to demand that such a medicine as alcohol should be brought before the bar of medical jurisprudence, and subjected to the rigorous severities of chemical criticism, and the not less close analysis and accumulated testimonies of an experience which must respond to the graphic method of close clinical delineation. All the more, because on this question is dependent the interests of society more than upon any article in the whole range of our science and art.

Most of the authors on Materia Medica, Practical Medicine, and Hygiene, recognize the general fact, and so interlard their books with moral cautions as to show some recognition of the degree to which society is involved in the medical status of alcohol. There are, therefore, double reasons for the most untiring study of alcohol in its bearing in chronic diseases.

Any medical man who has been in practice for the last quarter of a century, can not but recognize the wonderful change which has taken place. We have come to understand more thoroughly the laws of alimentation, and to see how much more can be accomplished by nutrients in cases in which stimulants were once the chief reliance. Ammonia, ether, effervescents, beef-tea, ginger-tea, coffee, and common tea, accomplish much, formerly entrusted to spirits. "The fashionable plan," says an advocate for their employment, "of giving great quantities of strong spirits, is happily dying out, and is being replaced by a more careful practice."

In the whole management of lung diseases, with the exception of the few who can always be relied upon to befriend alcohol, other remedies have largely superseded all spirituous liquors. Its employment in stomach disease, once so popular, gets no encouragement, from a careful examination of its local and constitutional effects, as separated from the water, sugar, and acids imbibed with it. Aitken's "Practice of Medicine" says of the alcoholics : "As an irritant, they stimulate the glandular secretions from the mucous membrane, and ultimately lead to permanent congestion of the vessels and to thickening of the gastric tissues."

The power alcohol has of drying the secretions, of congesting membranes, of complicating thé assimilation of food, save when largely diluted, is unsurpassed by any known remedy in general use. We do not say but that, by its sudden impression on the nerves, it may act as a counter-irritant to some present irritation or depression, or help to liberate confined gas. So will ammonia, distilled vinegar, warm ginger-tea, and various other medicines.

The liver, so important an organ of elimination in its relation to digestion, is among the first to suffer both from malt and fermented liquors. The change that takes place in the connective gland-tissues of the stomach and intestines, indeed, would point us to its destructive interference with glandular functions in advance of the degradation which occurs in the liver and kidneys. The assertion of Cameron's "Hygiene" (p. 282) is fully justified by the teachings of pathology, and the tests of exact trial.

"In candor it must be admitted that many eminent physicians deny the efficacy of alcohol in the treatment of any kind of disease, and some assert that it is worse than useless."

Most who, after expert and clinical study of it, claim for it a place as a medicine, confine it, to those cases of sudden failure of heart action which need relaxation of the circulatory vessels and such sudden nerve effect on the heart as will increase the vis a tergo or to those cases of adynamic prostration where it must necessarily be administered by an attendant in apportioned doses. It is, in fact, almost confined to the sphere of a diffusible stimulant acting by a paralyzing effect on the nerves of the blood-vessels and by an irritation of the nerve structure rather than by any physiological process of stimulation. We claim to have examined with some degree of diligence the testimony of others as set forth in books and in articles of medical journals, and to have observed carefully for ourselves with the honest intent of arriving at the true therapeutical effect of alcohol. Abundance of opinion can be found both in adulation and condemnation, but it is very noticeable that those who feel themselves bound to speak in accord with the physiology of function, and the pathology of organic changes, and those whose experience is such as has deprived clinical observation of that looseness which must occur where alcohol is administered amid manifold nutrients and medicines, are expressing their views with far more doubt as to the efficiency of alcoholics.

We find ourselves compelled to say that the same kind of study which has discredited the use of calomel as a special cholagogue also abates all alcoholics as digestants. The same methods which authenticate starch, and sugars, and oils, and milk as foods, compel us to omit ardent spirits from the enumeration. The explanations which show the value of water as a universal menstruum, of the various condiments as hydro-carbons, of tea and coffeé as nitrogenized exhilarants, will not suffice when we attempt to make of alcoholics a kind of food-medicine.

So with most of the suggestions as to the agency of alcoholics in disease. We can see how the normal expenditure of force increases progressive metamorphosis. We can meet the demand not by giving doubtful liquids, but by such foods as by quick digestion or rapid assimilation supply the want. In disease we are fast finding out that the demand created by accelerated waste is of the same nature, and that such foods become the real medicines. If change of tissue goes on so rapidly as to exceed the limits of conservative waste, we may also begin to study how to make the given amount of food go further, or to make up for weakened heart-action. But this is to be done by sedation, by the recumbent posture, by reduction of temperature through unquestionable methods, rather than by the use of a medicine claimed to retard change in an inexplicable way, and known in its physiological action to entail functional defects and organic lesions such as never attach to water, quinine, phosphorous food, or to various direct sedatives.

PURITY OF ALCOHOLICS.

There is another point in reference to alcoholic liquors as a medicine which, to a great degree, nullifies their applicability, while it throws the gravest suspicion upon the credence to be given to much of the clinical experience adduced in favor of alcohol as a medicine.

If we are to consider any article medicinally, we ought to be able to know that it is the article it professes to be. The clinician who claims uniformity of result and a declared therapeutic effect from a given article, has reason to question whether the effect was a result of the conceived cause when he finds that the cause was necessarily inoperative by reason of absence or admixture. If the article used was not tested either as to quality or quantity; if the general rule in respect to it is unreliability, we surely can not trust general opinion as to its dietetic or medicinal value.

We are used to hearing criticism as to the composition of liquors without much professional impression therefrom, supposing that these criticisms were chiefly based upon general accusation, or the trial of bar-room specimens, and that the bonded warehouse and the druggist's label were assurance of purity. But in the medicinal examination of liquors we find some facts which address themselves to us most directly in their clinical bearings. In our inquiry in this line we shall pursue the same course as when examining into food-value, seeking our information from chemists, from medical experts, and from the best accepted authorities on the composition of alcoholic medicines.

I. *There is much variation in different spirituous liquors and in different specimens of the same kinds of liquor as to the proportion of alcohol and other ingredients, independent of any question as to adulteration.*

A reference to the Proportion Tables of Brande, Fontenelle, Bence, Jones, Thudicum, and Dupré, etc., will show at a glance such variations as these. Wines in general have from 8 to 26 per cent. of alcohol, viz.,

Madeira, from 19 to 26; Port, from 17 to 25; Sherry, from 19 to 25; various other wines from 8 to 16 per cent.

Thudicum and Dupré state that "a natural wine may contain from 9 to 16 per cent. of alcohol."

Fowne says of vintage wines: "The quantity of alcohol varies very much." Sherry wine often has from 6 to 8 gallons, and port, three gallons, and the best port 13 to 15 gallons of brandy added to the cask. This is not considered admixture, but original "fortification," as it is called. Native European wines, by the English law, must not have over 26 per cent. of alcohol. Champagne, Burgundy, and the Rhenish wines vary from 8 to 14 per cent. by measure of alcohol.

Such statements are made still more significant from the authority of Parkes, who says, "So various is the amount of alcohol in wines from the same district, that a very general notion only can be obtained by tables, and samples of the wine actually used must be analyzed" (p. 265).

Brandy, whisky, and gin vary in from 50 to 60 per cent. of alcohol, and rum from 60 to 70.

Apple-brandy should have from 40 to 46 per cent. of alcohol.

As we shall hereafter see, these are so varied in other regards in the various processes and methods of manufacture, that their alcoholic variation is an inadequate measure of their therapeutical import.

Cider varies from 5 to 10 per cent. In brewed medicines, as represented by ale, porter, beer, the alcohol

5

varies from 1 to 10 per cent. (Parkes, p. 173). In the London small-beer it is as low as 1.78 per cent.

Lager-beer varies in alcohol from 4 to 7 per cent. All ingredients of nutritive value in these can be furnished much more cheaply and more in accord with the ascertained principle of adequate assimilation and nutrition.

If, as claimed by most, the alcohol is the chief available medicinal ingredient of alcoholic stimulants, it is clinically requisite to know just how much we are administering in each dose prescribed. This all the more when leading advocates of alcoholic treatment now claim that the limit consistent with health is one and a half ounces in twenty-four hours, and that the limit in disease is far narrower than once believed. In face of the grave facts which come to us from pathology and from practice as to the effects of overdoses on the nervous and vaso-motor system on the circulation and upon secretion and excretion, it is *not good practice* to deal with this article without definite knowledge of the dose administered.

That must ever be an imperfect clinical study of any medicine which is unable to state the amount administered in any given case. A thorough examination of evidence compels us to say that we are prescribing alcoholics too empirically if we are using the articles commonly in the market for medicine without having had the specimen chemically examined and certified. *The rule is impurity or great variability in the alcoholic proportions*. We would not use as a sedative a mixture containing morphine, not knowing whether it had one grain or four to the ounce, if it was the effect of the morphine we were chiefly seeking. The testimony of chemists as to variability is so uniform as to be stultifying to all precision as to the usual article.

The practitioner has to watch closely effects, not merely to find out the adaptation of the remedy to the disease, but to find out how much he is giving (see Loomis as quoted, Parkes, Anstie, etc.) Surely it is asking too much of a science and art involving life and reaching out toward accuracy, to be laden with such sources of error. If it is a real medicine, and must be used, let us draw the line of demarcation between the manufacturers, the wine merchants, and the compounders, and get our medical supply from the chemists. The trade article is already marked, " Composition not known."

(a.) *The time is already at hand when in speaking of alcoholic liquors we must state which of the alcohols we mean, as well as how much of each is present.* These are often mingled, and their physiological and medicinal effects are totally different. Let us gather up briefly from the chemists and experimenters, such as Fowne, Attfield, Richardson, Prescott, Parkes, a few significant facts which are, we believe, not gainsaid, and which present the question of the medical use of alcohol, so-called, in quite a new light.

Fowne gives a list of twelve alcohols as belonging to what is known as the monatomic series. Of these, five chiefly concern us in medicinal and dietetic relations. These are the

Me-thy-lic (wood-spirit) . CH_4O.
Eth-y-lic (or common), . C_2H_6O.

Pro-pylic, ...	$C_3 H_8 O.$
Butylic, ...	$C_4 H_{10} O.$
Amylic (grain-spirit), (fusel),	$C_5 H_{12} O.$

Fowne says the term alcohol, originally limited to one substance, viz., spirit of wine, is now applied to a large number of organic compounds, many of which, in their external characters, exhibit but little resemblance to ordinary alcohol. Yet, as we shall see, no usual liquor is clear of some one of them besides the ethylic, which is the one known as common alcohol, and secondly, their effects are medicinally quite different.

The first of the series methyl alcohol, naptha or wood-spirit, is not an original impurity in manufacture of an ethyl (or common) alcoholic, but in the form of "methylated spirit it has been added in the falsification of liquors" (see Prescott, *Prov. Journal of Medicine*, Detroit, April, 1876). Pure methyl alcohol is a colorless, thin liquid, *very similar in smell and taste* to ethyl (common) alcohol (Fowne, p. 570), and like it, is a mobile, watery liquid. "It has an aromatic odor, free from empyreuma, and a sharp, but not acrid taste" (P., p. 262).

Richardson (see Cantor Lectures on Alcohol, p. 30, and also Munroe, "Is alcohol necessary?") has made some important inquiries in respect to it. He names the case of a well-known physician who himself used it in preference to common alcohol, regarding it as less injurious. Richardson has long used it in his practice as a substitute for the ethylic or common alcohol. He says, "It is much more rapid in action, and much less prolonged in its effects, than is common alcohol, so that it produces its effects promptly, and what is of most importance, it demands the least possible ultimate expenditure of animal force for its elimination from the body. This latter fact is of great moment, for, in the end, all these alcoholic fluids are depressants, and although at first, by their calling vigorously into play the natural forces, they seem to excite, and are therefore called stimulants; they themselves supply no force." He notices its special power of reducing the temperature (p. 32) and calls it the lightest and least injurious of the alcohols (p. 33). If we are giving àn alcoholic medicine having this in it, we are using the spirit most allied and yet very distinct from ethylic alcohol.

The propylic alcohol is not much understood in its medicinal action. Chancel (F., p. 531) obtained it from fusel or amylic alcohol by fractional distillation. One of its isomeric modifications has a very close resemblance to ethyl alcohol (F., p. 531). It is not unfrequently found with it in liquors, and of course modifies its characteristics. Its medicinal action is not yet well enough understood to mark the contrast between it and the ethyl drinks or medicines. It is, if possible, even more greedy of water, and so still more robs the tissues (see Fowne).

Butylic alcohol is often associate with amylic alcohol, and so is present in the usual fermentations. When valerianic acid is prepared as generally artificially from butylic alcohol, the various valerianates are often contaminated by butyrates (see Atthill, p. 322). There is reason, therefore, to believe it often present with the other alcohols. Richardson seems to have

shown that this heavier alcohol is much more apt to
cause tremors of that kind occurring in delirium
tremens.   Its effect is certainly very different from
that of the ethyl alcohol for common medicinal use,
and must be studied by physicians as to its own distinct effects.

The next in the series is known variously as amylic
alcohol, potato oil, grain oil or grain spirit, and fusel
oil.   In its relation to ethylic, or the assumed common
alcohol of alcoholic liquors, it is of most interest to
the physician.

"It constitutes (says A. B. Prescott, 1876, *Prov.
Journal of Medicine*) the chief part of fusel oil, which
also contains the fourth alcohol (butylic), and traces
of the third, propylic, and these together appear quite
invariably in the alcoholic fermentation of sugar—
their proportion being varied by conditions."   "*It is
the great and naturally occurring impurity of liquors*,"
says Fowne, p. 535.   "In the manufacture of brandy
from corn, potatoes, or the must of grapes, the ethyl
or common alcohol is found to be accompanied by an
acrid oily liquid called fusel oil, which it is very difficult to separate completely from the ethyl alcohol."
When we are shown some of the fatty acids which
are extracted from alcoholics in very small quantities,
we are not to conclude that this is all the amount
that can be found in our usual liquors.   Common
alcohol dissolves it readily (see Fowne); when diluted
it has a sweetish, fruity taste and smell, is not over
pungent, and is easily relished.

Attfield (" Medical and Pharmaceutical Chemistry,"
p. 389) says, "Amylic alcohol is a constant accompani-

ment of ethylic or common alcohol when the latter is prepared from sugar which has been derived from starch." This, of course, applies to all alcoholics prepared from grains or vegetables either by brewing or distillation. Not only does it thus occur naturally, but there is reason to believe it is often added. Nor does the seal of the bonded warehouse protect us from various admixtures that may occur before importation. This is the most injurious of the five alcohols thus far noticed. When produced as so often in manufacture, or when added for profit and improvement of flavor, it is "an extremely dangerous addition to ordinary alcohol." Its action is the same as that of butylic alcohol, "except more prolonged by reason of its greater weight and insolubility" (see R., p. 37). It is a usual and a profitable ingredient.

In this country, where whisky and other fermented products are so largely used, the art of their concealment and modification so as to give a relishing flavor is well understood. There is little doubt but that certain effects of alcoholic liquors of late years are due to the more profitable utilization of the "contemporaneous" alcohols with the ethylic.

It is also true that the addition of such alcoholics as the methylic, or of various substitutes for all the alcohols, has given to what is called the experience of prescribers a character somewhat mythical.

In the various modes of preparation and manufacture, variations occur not only from different vintages, distilleries, and breweries, but from difference in method, even within the limit of legitimate preparation. Modern chemistry shows how extended is this alcoholic series. Although "analogously constituted, they are of very dissimilar effect and variably mingled in manufacture" (Fowne). "Fusel oil," says A. B. Prescott, in his recent excellent book on " The Chemical Examination of Alcoholic Liquors," "can be found in the larger number of liquors" (p. 17). Some wines have it in manufacture still more plentifully in wine from the marc of grapes, *i. e.*, in wine made from the must of expressed grapes. This, under certain management, yields much wine (see J. J. Griffin, Chemical Testing of Wine and Spirits, 1872). Whisky is often so distilled as to allow much amylic alcohol or fusel oil. As dissolved or mingled with the other alcohol, it gives that taste so much relished by systematic drinkers.

Intoxication and the manifold effects of alcoholic medicines nowadays are not solely the effect of ethylic alcohol, but a complex result of various toxics on the nervous system and the brain. And we, as physicians, are so little familiar with the effects of ethylic alcohol in itself considered, that we have little reason for positive clinical deductions unless the specimen in use has been chemically examined and certified.

In spite of care, there will be differences, and the art of manufacturers can change the product somewhat, as does the milkman who swill-feeds his cows, but has too much conscience to add diluents afterwards. If this is skillfully done, it does not produce results which can be identified as deteriorations, but adds largely to the profits. A distinguished analytical chemist of our own State recently drew our attention to the degree to which the art of the distiller now made

it profitable for other ingredients to distill over. Where
the standard of purity is so complicated and the adroit
push and advertising of the manufacturer and the rel-
ish of the steady drinker are the marketable tests, it
is no wonder that there is such lack of uniformity.
No one will examine, as we have, the circulars of dis-
tillers or the more exact directions of the leading
authorities on liquor-making, without being satisfied
as to the great variety of compounds merchandized
under one general name.

This variableness of veritable alcoholics as resultant from variations of original distillation has been
too much lost sight of by us in our medical use of
the various preparations. There are still other variations arising from variableness in other constituents.
Brandy contains some ether, fusel oil, tannin, etc.
Whisky, as defined by A. B. Prescott, in his
"Chemical Examination of Alcoholic Liquors, 1875,"
is the diluted alcohol distilled from fermented
grain (malted or not) or potatoes (p. 718). Wine, according to Pavy, is the result of the blending of many
constituents, and from ten to fifteen distinct ethers
can be traced in some specimens. What is called the
bouquet of wines is largely dependent upon various
ethers or essential oils almost as variable as the specimens produced. The same is true of the sugar, extractive matters, and salts. Beer consists of malt and
hop extract, sugar, and free acids, salts, and starchy
matters, and potassium, magnesium, and phosphates.
Ale and porter have a share of the same. All
these have a very wide range of proportion (see
Parkes, p. 257, Fowne, etc.) Cider and perry as well

5*

as wines, being made from fruits, have sugar, acids, and salts, and so have some of the effects of these.

All of the distilled spirits are greatly modified in their effects by the degree of dilution. These two differ through a large range in solids and acidity (see Mulder and Hassall, p. 269). The good effect of some wine in very low conditions of fever we believe depends far more on ethers, essential oils, and other ingredients than upon the alcohol.

(*b.*) Great modifications of alcoholic liquors occur from age. This is often difficult to ascertain with reliability from dealers or the tests of chemical examiners. Indeed, some of the characters which attach to age are modified by mode and place of keeping. It is not merely that the flavor or bouquet of liquors is altered, but the characters and proportions of ingredients are modified. The chemical examination shows difference in the amount of alcoholics, and in various extractives and ethers. Apple-brandy is never fit for use until three or four years of age, and the best professional judge we know says it should not be used medicinally until six or eight years old.

Much of the whisky sold, says Cameron, of Dublin, is unfit for use on account of its freshness. When whisky is first prepared, it contains certain matters, especially fusel oil (amylic alcohol), and, if removed by rectification, the whisky is converted into spirits of wine, which, though pure, is unpalatable.

Such terms as wine, brandy, whisky, spirits, are as indefinite in our Materia Medica as are the terms liver complaint, fever, and dyspepsia in disease. We are revising our nomenclature of the one, and must, in like manner, make our alcoholic nomenclature to designate some definite article rather than to indicate merely analogous compounds. Indeed, one can not institute an honest medical examination of the facts as to alcoholic liquors within the bounds of what is generally assumed as purity, without feeling that the prescription of an alcoholic liquor with the view of obtaining results supposed to inhere in alcohol, lacks both scientific and practical accuracy unless the article used has been duly tested.

There is about it an inexactness which would attach to the use of an indefinite amount of opiate in some unknown mixture. In such cases we can not but be in doubt how we are prescribing, or what proportion of effect to attribute to each item in the compound.

It is far more in accord with the physiology of digestion, and with what we know of animal chemistry, to attribute some of the effects believed to be produced by alcoholics to other ingredients than the ethyl alcohol they contain.

Sugar, starch, acids, hydro-carbons in the form of oils, extractive matter and bland water in liquors, are not necessarily deprived of their dietetic and medicinal qualities, because associated with other ingredients, either useless or harmful. The various ethers found in most liquors have some real exhilarant value yet to be more thoroughly apprehended. If such is the case, the question to be determined is how best to give these, or whether the harmful accompaniments of the alcoholic series can not be separated.

In a careful study of the variations in different liquors, and in those of the same name, it seems marvelous to us that they are so generally associated together in their therapeutical relation.

A recent excellent Materia Medica, after noting, with some precision, the declared value of alcoholics in enumerated diseases, and stating their value and uses, in the summing up, says, so far as spirits are concerned, " I have never been able to perceive any difference in their action (gin, of course, being excepted), save only that sometimes one agrees better with the stomach than the other. This has seemed to me to depend simply upon the personal likings of the patient, to which, therefore, the choice may well be left."

How can we label that " experience ? " Dr. Munroe, F.L.S., rightly asks : " On what grounds can the scientific physician order his patient to take daily quantities of rum, brandy, gin, or wines obtained from publicans or dealers, when he can, without analysis, have no knowledge of that which is prescribed, or the effects that will be produced ? May not the promiscuous administration of these intoxicating drinks be pronounced to be highly empirical ? "

Dr. Thos. P. Lucas, in his article on " The Physiological and Chemical Action of Alcohol," (*British Med. Journal*, 1876, II., p. 612), says : " To give brandy and port, is to give we know not what. What would be thought of a practitioner ordering his patients to take a few tumblers of sea-water whenever he wished to give iodide of potassium, etc. Yet such is the practice with regard to alcohol. We give it in all diseases and in unknown strength."

In dealing with an article which confessedly makes its chief records in the nerve-centres of life, and

whose ill effects are registered with a severity that attaches to no food or freely prescribed medicine in all
the materia medica, are we not called upon to get
upon a different basis instead of associating a vast
body of toxics, and claiming results as from one, when
really there has been no uniformity in prescription?

If a physician could write out the details of
every alcoholic prescription made the last year,
the pointings of science and of actual examination are, that, instead of a few general names, we
should have, choke-full, a vast prescription-book of a
farrago. A medley such as no other article of the
materia medica could match.

Dr. J. C. Morris, in a recent paper, gives some details as to the different effects he has noticed in the
effects of beer and light wines upon the populace on
festive days, as compared with that of distilled liquors.
The extractive matter, starch, sugar, and hops of many
of the malted drinks, can not but be quite different
from that of alcohol, as it appears, for instance, in apple
whisky. He refers to the remark of Thudicum, that
the effects of alcohol should be carefully distinguished
from that of aenanthic ether, etc. Morris advocates
the lighter wines to stimulate the cerebro-spinal system, "in which the aenanthic ethers produce a large
proportion of their effect." Yet he names all the distilled liquors in a compacted list, as do most. Only
by accurate and ascertained distinctions, shall we ever
be able to know the value of liquors as medicines.

II. *Next, it is germane to our subject to inquire
whether outside of these variations in original preparation and permitted difference of method, there are*

others which accrue from direct substitutions and adulterations, and to know how far our confidence in use and in clinical experience may be thus modified.

As we come to the consideration of this department, the sufficient uncertainties before recognized as to the qualitative and quantitative composition of the various alcoholic medicines, is quite indefinitely multiplied.

Wines, according to such standard authorities as Thudicum and Dupré, Mulder, Parkes, and Richardson, are varied by water, cider (P., 268), ethers of various kinds, aldehyde (R., p. 36), logwood, tannin, and other astringents. Copper, alum, lead, lime, and salts in excess are often detected in specimens. For instance, Prof. Parkes says : " Port wine as sold in the market, is stated to be a mixture of true port, marsala, bordeaux, and cape wine with brandy. Inferior kinds are still more highly adulterated with logwood, elderberries, catechu, prune juice, and a little sandle-wood and alum. Receipts are given in books for all sorts of imitation wines. Battershall, in his recent book on " Legal Chemistry," 1876, says : " The common adulterations of wine with water is somewhat difficult to detect " (p. 136), and refers to " toxical substances as often contained in wine " (p. 141). There are (see Prescott's Examinations) fictitious wines of all grades, and some have not over two gallons of wine to a barrel.

Brandy, says Cameron (" Manual of Hygiene," p. 322,1875), is often adulterated with corn-spirit (whisky), and therefore frequently contains fusel oil. " It is generally mixed with a little syrup." It also often contains tannin, and coloring matter, and ethers of various kinds (P., p. 269), besides oils which come over at the close of distillation.

The various beers are very uneven in their composition. In examined specimens there have been found sufficient of cocculus indicus, salt, copperas, opium, Indian hemp, strychnine, tobacco, darnel seed, logwood, salts of zinc, and lead and alum, to lead the English License Act of 1772 to single out these from other adulterations. The English ales and beers are less treated with injurious additions than formerly, but are not more uniform or less replaced by substitutions such as chemistry provides. "Many of the high-priced beers contain now little else than alcohol and bitter extract." Parkes' "Practical Hygiene" (1873, p. 286) says: "Seldom are the adulterations more injurious than the alcohol itself; but by a super-abundance of phosphates, hydro-carbons and sugars, fatty changes are often induced in the liver and kidneys, and too great diuresis promoted. Increase of fat under such circumstances, as a result of excessive malt foods, and undue stimulation, has very narrow limits of advocacy in comparison with other methods. In his speech in the House of Lords, April, 1872, Lord Kimberly says a common adulteration is as follows: "A certain amount of beer is drawn from the cask of 84 gallons, and then 6 pounds of 'foots' (a black, coarse sugar) and 1½ gallons of 'finings' (made from skins of soles and other fish), and 12 gallons of water, are put in per cask."

Although brewed liquors undoubtedly contain some of the materials which make food, and may act medicinally, yet they are in unfortunate company, and can be supplied to the system in a much more phar- maceutical and economical way. In these modes of combination there is physical as well as moral hazard. The physician, too, needs not to lose sight of the fact that they are the most expensive of foods or medi- cines to the classes for whom they are most pre- scribed.

Whisky, as a drink for the common people, has invited the attention of manufacturers so fully, that they have learned many arts for direct adulteration. The whisky frauds are not chiefly those of St. Louis, and the whisky ring has other branches of knavery than that of cheating the Government. We may blush as citizens, but as physicians we can stand the political imbroglio. Yet as physicians we can not stand it, that to recover faint hearts and to rouse ex- hausted nerve-centers in adynamic disease, we must go to medicated liquids, so unreliable as are most of our alcoholics.

Rum is often mingled with molasses, and poorly distilled. It contains a good deal of the injurious bu- tyric ether, to which it owes much of its flavor.

Gin is so far manufactured as to have no test stand- ard. What we call gin is chiefly a medicated diuret- ic, in which common spirits, juniper, and turpentine bear an important part.

President Barnard, in his article on the " First Cen- tury of the Republic " (Jan., 1876, *Harper's Monthly*), says: " Of the different descriptions of strong liquors, of which, to the misfortune of mankind, so incredible quantities are annually consumed as beverages under the names of rum, gin, choice brandies, superior old Bourbon, Monongahela, etc., probably half or more than half of the quantities sold are merely diluted solutions of alcohol, to which chemically-prepared essential oils, and chemically-prepared sugars, have communicated so perfectly, the odors, flavors, and colors of the liquor imitated as to defy detection by the most practical dealer or drinker."

III. *There is still another most important therapeutical view to be taken of the various alcoholics.* The time has already come when prepared liquors, which have never seen a vintage or a distillery, claim their right to be accounted genuine and not to be branded with the epithet, " adulteration."

The ablest and most candid French work and authority on the manufacture of liquors (Duplais) discusses the question from the stand-point of business integrity, and leaves it without adverse decision.

Prof. A. B. Prescott, the able chemist of Michigan University, in his work on the chemical examination of alcoholic liquor, says: " The term brandy, as used in commerce, without qualification, must be held by common consent to include artificial brandy." He makes three divisions, viz.: veritable brandy; artificial brandy, or that made of the same ingredients, according to chemistry; and third, fictitious brandy, *i. e.*, a fraudulent imitation. There are many preparations known as brandy essences, which are so combined with spirit as to make artificial brandies difficult of distinction from the veritable (see Prescott).

Griffin, who is authority, and has examined many

specimens, says that he found what he called public-house brandy not to be a distilled spirit, and yet it had the taste of a good Cognac. He gives a German recipe, which, he says, after long keeping, gives a brandy very similar to real Cognac in taste and odor. He adds that this sort of sophistication is evidently becoming an ordinary commercial practice. The recipe, as given by him, contains much purified spirit, but only 8 quarts of wine to 150 quarts of mixture.

Duplais' most able and scientific work of 700 pages, octavo, translated by McKennie, and published by Baird, Philadelphia, 1871, would be incomplete without its pages on "Imitations." Speaking of Cognac, the most difficult of all brandies to prepare artificially, it says (p. 293), "It is a fact worthy of note, that the brandy obtained in consequence of the addition of a spirit foreign to the wine in limited proportions can not be distinguished from the natural wine by itself, that is to say, without the addition. Finally, brandy resulting from the new method defies all methods of investigation. We may suspect the mixture, and even know of its existence, but we can not furnish the proof; neither the most skillful and practiced taste, nor the persevering researches of the most skillful and learned chemists have been able to detect it. M. Payen himself acknowledged some time since that in the actual state of the science, the discovery of the mixture presents insurmountable obstacles."

Such a book as that of Pierre Lacour on "The Manufacture of Liquors, Wines, and Cordials without the aid of Distillation," and others well known to makers, and but little to physicians and pharmacists, show to what perfection the experts have attained.

The chapter of· Duplais on the preparation of various liquors, of which alcohol, sugar, and water are the basis, also shows how artificial and successful are such compounds. The author, after admitting that "the *immoderate* use of spirits and even liquors is pernicious," and after designating the drunkard as "a wretch who is unworthy to live" (pp. 437--8), claims that the "aromatics and sugar are good for digestion," and the small quantity of alcohol not injurious to persons in health.

We have had occasion to examine recent popular books of recipes for the imitation of liquors, and it is wonderful to what perfection the art has been elaborated.

When we look into our recent chemistries and see how many natural products are duplicated in art, and that fruit, essences, and flavors are so well imitated that connoisseurs can not distinguish the artificial from the real, it is to be expected that the great profits accruing would induce the production of liquors without the trouble of vintage, wine-press, or distillery.

More Madeira wine is on the market than could be made on the island of Madeira if each grape made a gallon. Critical examinations have gone far enough to show that in all alcoholic drinks as consumed for thirst or medicine the veritable article is largely reduced or superseded by admixtures and imitations.

The editor of the *London Lancet* (Nov. 1, 1873) well says, "We suggest it were well worth our while in the future for us to transfer our allegiance when we would

prescribe alcohol from the wine and spirit merchant to
the chemist " (p. 639).

*When imitations are repudiating the reflection of
being "spurious"* and claiming legitimate place beside the veritable products, and when from either the
" fictitious " is difficult of distinguishment, is it not
high time that the therapeutist or physician take
reckoning as to his *professional* position in reference
to alcoholics as at present prescribed ?

We commenced the examination of the dietetic
and medical value of alcoholic drinks, we confess, with
only moderate misgivings as to their availability.
We have been accustomed to their use under the in-
struction of authors, although in the line of experience
we had come to doubt their utility in very many cases.

Bound by no pledges of personal abstinence, and
feeling personal responsibility to use the best remedy
which the case in hand might require, we have conducted this study as we would that of any article.in
the armamentarium of foods or medicines, with the
one exception that we have confined ourselves mostly
to chemists and practitioners, and even slighted the
evidence of physicians who as temperance reformers
had been accused of pre-judgment.   Careful examinations of scores of testimonies from chemical and
medical experts show that from any such tests as the
natural sciences institute, alcohol has thus far failed
to establish itself as a contributor to tissue or to
force in such way as to entitle it to rank with foods.
This has not been owing to the lack of diligence on the
part of those who would so classify it.   Personal and
popular views can not but give an unconscious zeal

and bias of opinion and lead observers often to be too easily convinced in the line of their own preconceived notions. But it is a study by itself to notice how, one after another, many of the dietetic and curative claims for alcohol have vanished.

When we come to study it in its direct medicinal relation to pathological conditions, and subject it to those methods of test to which we subject quinine, iodide of potassium, etc., we find its *modus operandi* embarrassingly obscure, and its results negative or positively injurious except within the narrowest limits.

As a sedative and narcotic, its effect is so indirect, and except in toxic doses so inconsiderable, that few are now found to place it in this regard beside the recognized class of real sedatives.

As a digestant it is so nearly reduced to a local excitant, and is so apt to be associated with accessory and positive evils, that it can not be urged here, by the side of many other remedies now available to the profession. Its effects upon the circulatory system seem to be a part of its action on the nervous system. This disturbance of innervation is so pronounced, that we readily name it " Intoxication " or " poison-work." It is within very narrow limits that we should attempt to regulate life-currents by such an agent. Indeed, attention is concentrated by the index finger of imperative facts, upon its special and toxic action on the two great nerve systems. And when we read and see its declarative effects in what is called excess, and its dubious advantage in the estimation of skillful experimenters, we must feel it to lie under the gravest suspicion as a remedy.

Animal chemistry, histology, and physiology can not establish for it any normal occupation, while pathology and disordered function weary over the demonstrations of its manifold impediments. We then flee for rebutting evidence to clinical experiences.

If we could find agreement here, we should still hope to find science enlightened and guided by art. But here we have not only want of conclusive evidence, but reason to question the records of many so-called experiences. In respect to no article in the whole range of medicines are there such doubts as to quality and quantity given, as to the agency of different components of the mixtures used, and as to the real benefits secured.

We are compelled to accept the summation given in the *Blythe and Tardieu Dictionary of Public Health,* 1876, in an able article which says: " The truth really is, that it has been prescribed, even by the most eminent men, under the forms of beer, wine, spirits, the strength, adulterations and combinations of which are seldom in any given sample known, in the most opposite affections, and as a result, it has on the one hand been extravagantly given and lauded to a most unwarrantable degree, while on the other hand, by another class of observers it has been entirely withheld " (p. 37).

Whatever may have been the preconceived views of physicians, we believe any one who will candidly sit down and study the leading authorities on materia medica, and therapeutics, and the practice of medicine, and by their side study the details of chemical physiological investigation, and the methods of general medicinal use, will be satisfied that the therapy of alcoholics needs to be carefully reviewed from a strictly dietetic and medicinal stand-point.

While on the one hand it is alleged that the prejudices of reformers in the interests of abstinence may lead them to extremes, we have in various ways evidence that traditional and popular beliefs, and the forces of habit and authorized practice handed down by mere weight of general authority, need careful reviewing by the light of those more exact methods of test and of observation, which are now the aim and tendency both of our science and art.

If we are to shut out the testimony of the devotees of total abstinence on an assumption of bias, we must also shut out all those who themselves use alcoholic drinks in any form, and leave the unprejudiced investigation of the question to those physicians who are not identified with temperance movements on the one hand, and on the other are not under the unsuspected influence of that prejudice which the self-joyment of a daily glass of wine is equally apt to induce.

It will not do for us to talk pitifully of the sufferings of a cause "from the intemperate zeal of its advocates," while they with equal justice can talk to us of an advocacy sustained more by our personal temperate use than by any deductions of chemistry or therapeutics which bear the test of accuracy. Our very craft is compromised in that, in the absence of such value as such an article should be able to authenticate from the testimonies of science or of art, we are using it most freely as a medicine in a way too empirical for any experimental science and too un-supported by careful *experience*.

We are of those who do not deny that alcoholics are luxuries. Wines, champagne, and some of the distilled liquors, nicely prepared with sugars, flavors, etc., do taste good. Men constantly indulge in the use of some things which are neither useful as foods nor medicines, but which, nevertheless, they enjoy. We have always admired the frankness and judgment of a distinguished Philadelphia professor who, with force more profane than elegant, used to say that good Cognac was no food and poor medicine, but a drink that he very much liked.

Because as such it does not in small doses always make palpable record of its incompatibility, or be-cause the man who has taken so much as to become toxic with joy seems to return afterward to his wonted vigor, we are not therefore to minimize its evils or wonder if somehow it has not vindicated for itself a medicinal function. An article may be taken into the system which does not accord with any physiological or reparative demand, and yet the body be able to re-bound therefrom just as it makes thorough recovery from over-fatigue and from manifold abnormal things. Nay, such are the recuperative powers of nature, that it is not possible to say that in every case life is always shortened or permanently reduced in force by indulgences which are neither normal in health nor remedial in disease.

We may even go further, and say that by the use of a food or medicine, a toleration of the same may be established so that the evils resultant therefrom

are not at once such as the legitimate consequences
would entail. *This is so marked as to tobacco* that
Bintz says that persons inured thereto are not to be
taken as the criterion of its effects (see article on
"Some Effects of Alcohol"). There are balancings
and adjustments of our physical nature within certain restricted limits that do not enable us to say
of an article not required that it is never tolerated
or that it registers in every person and in the same
degree the harmful nature of the article itself. There
are two factors, the article and the body using it.
The evil effects of the one may be modified by the
special conditions of the other. The body is more
tolerant of an evil at one time than another. Some
bodies are more tolerant of an evil than others. A
body may adjust itself to food substitutes within
certain limits, and being deprived of its natural aliments, and the organs for their assimilation becoming
changed by disease, may come to distill and extract
force as does the horse from a whip or in some other
way. But such assistance or support is neither
natural or to be generally sought.

Our corporeal nature is compelled to put up with
deleterious *substances* introduced into the lungs and
the stomach to a degree we as physicians have never
yet realized. Its marvelous adjustments within limits
endure these, shake off results, and the bow springs
back with its tension not permanently enfeebled to
the degree that mathematical calculation might indicate. There is a correlation of forces within the
animal economy as well as in outer nature. But this
must not be construed into making alcohol a food or

6

too widely a medicine.   These facts must not be
whipped into evidence that therefore alcohol is
sometimes necessary. We must not proceed from
such data to call it indispensable, and to indulge in
dicta totally unsupported in any way.

If alcohol could even be strained, as it can not, into
a food, or, as it can be, into an admissible medicine,
we would need to remember the sentiment of a distinguished chemist, that there are aids so ques-
tionable as only to be used as a last resort. If we
as physicians only confined all our alcoholic prescriptions to those cases in which no other article would
answer, I am sure we could constrict its use even as
a medicine.

We agree with the remark of Austin Flint in his
address on Practical Medicine (1875) before the
American Medical Association, " that the value of
alcohol as a remedial agent is to be considered inde-
pendently of the subject of intemperance," if by this
is meant that any such ascertained value is not to be
concealed or ignored in the interests of any reform.

But we think he would agree with us in saying that
the weighty facts as to moral as well as physical effects
which portray themselves in reference to this article,
make it all the more obligatory that we should call
for exact and discriminating evidence and guard
against a mistaken necessity in a case where mistakes
are so freighted with disaster. Responsibility of
opinion and of the exercise of an art has some relation to the immensity of the interests involved.   It is
as *professional* as it is humanitarian to demand that
in the study of such an article we should not commit

ourselves to loose opinions, or to judgments which
have none of the precision of clinical observation such
as will bear scrutiny. The sanitarian and the moralist
have a right to ask the grounds of our belief and of
our practice with the rigidness of severe scientific
scrutiny and of close clinical record.

Our preconceived notions must not lead us to attribute to alcoholics properties which neither science
nor art can prove. Unable to ascertain food-value or
medical-value according to the usual rules of evidence, we must not imagine evidence. If the article
used were inert, the case would be different and the
assertion of food or medicinal-values of little practical
import. But the opposite is too severely and in-
tensely true as to the material in hand.

The capacity of the alcoholics for impairment of
functions and the initiation and promotion of organic
lesions in vital parts is unsurpassed by any record in
the whole range of medicine. The facts as to this are
so indisputable, and so far granted by the profession,
as to be no longer debatable. Changes in stomach
and liver, in kidneys and lungs, in the blood-vessels
to the minutest capillary, and in the blood to the
smallest red and white blood disc disturbances of secretion, fibroid and fatty degenerations in almost every
organ, impairment of muscular power, impressions so
profound on both nervous systems as to be often toxic
—these and such as these are the oft manifested
results. And these are not confined to those called
intemperate.

We are aware that "*when used in excess*" is the
cylinder escapement for all this, but with facts drawn

not from intemperance, but so-called moderate use, with the facts as to the physiological effect on man and animals of doses not toxic; with its tendency to evil, and only evil, and that continually, the burden of proof of its medicinal value lies entirely with those who advocate it. When practical medicine tells us that three-quarters of all diseases in adults who drink at all, are caused thereby, and when pathology shows "its greatly predominating action to be that of a neurotic," even where its general effects are not so obvious, we may well watch it with a careful eye. The practitioner needs to respond with incontestable evidence of a value which rigid science pronounces untenable, and which variable experience has not established.

When the wailing cry of evil to society reaches us, high moral obligations require us to make out a clear necessity of use, or else ignore the article. If there is doubt, society in this case is entitled to the benefit of the doubt.

The medical, the pathological, the social, the moral questions are so imminent and urgent, so critical and crucial, that it is right to put each physician in the witness-box, and let him tell how he knows that alcohol is ever a food or ever a remedy. It is right to confront him with the results of manifold experiments, with the facts of skilled observers, with its failure to respond to the tests which estimate real food, and its inability to define its precise sphere as a medicine. It is right to bring out the physical evil that it does and may do, and show how this leviathan can be kept in its place. The physician should be able to say, " Hith-

erto shalt thou come, but no further." It is right to
question the reports of so-called experience, unless
the specimen used had been tested; unless he
knows the results depended alone upon *ethyl* alcohol,
and not upon other ingredients of the liquor used.

In face of facts such as we have adduced, it is a
burlesque on the precision of modern therapeutics to
deliver *ex cathedra* opinions as to the medicinal value
of alcohol, when we are using all sorts of chemicals
under that name, are attributing to them results which
are shown to be ill-sustained, and prescribing them
with an unscientific laxity quite inexcusable.

In the light of accumulating facts as to the whole
alcohol series, as to the manifold and variable combination prescribed under the cognomen of alcoholics, and with candid regard to the wide and contradictory range of their composition, we need to
review the records of our Materia Medica. We must
recognize the unscientific and empirical data on
which we have rested, as insufficient to overcome
the facts of histological and pathological research,
and feel that demand is made upon us to show the
conservative *modus operandi* of a medicine which so
significantly asserts itself as a toxic and depressant.

Such questions as these are up for exact consideration to a degree that neither science nor practice can
properly ignore:

What is the effect of *ethyl* alcohol as a medicine?

What trustworthy experiments have we as to its
physiological effects, and as to its therapeutical
indications?

What is the guide as to the quantity to be given?

What are the testimonies of experience as to its clinical use?

With present light as to the manifold variations of alcohols in legitimate manufacture, and its change by admixture, are the usual testimonies as to the advantage of ethyl alcohol as a medicine of any professional value unless composition had been determined?

How far can methyl alcohol take the place of ethyl alcohol as a medicine?*

To what extent does amyl alcohol or fusel oil occur in various alcoholics, either by virtue of manufacture or admixture?

How far is the effect of wines dependent upon their numerous exhilarating ethers, their albuminous matters, their sugars, acids, and salts; and how far is the effect of each kind modified by the varying proportions of these?

How far is the effect of whisky due to its ethers, essential oils, acids, sugars, malt, etc., and so as to all distilled liquors from grains?

How far does the effect of ales, porter, beers, etc., depend on sugar, starch, etc.?

How much are clinical experiments worth which speak of the value of alcohol as a medicine, with utter disregard of such matters?

---

* Richardson says, that it is practicable to substitute it. Prof. A. B. Prescott expresses the opinion that the views of Richardson are correct. The introduction of methylated spirit in place of the excise proof-spirit, has been a most important measure in Great Britain. (See "Report" of Prof. Graham, Hofmann and Redwood, Vol. VIII. of *Quarterly Journal of Chemical Society*, and Fowne's "Chemistry," p. 518).

Is it possible to deprive wines, etc., of all alcohol so as to secure their action independent of the spirits?

In the absence of chemical, physiological, and sustained clinical evidence, social science and moral considerations are confronting medical science and art, and rightly demanding of it the most incisive and convincing facts. If a medicinal use can be established for it at all, the same science which shows its use is incomplete even in this respect unless it defines limits and quantities and accurately designates the circumstances of adaptation. It will not do for us to express commiseration of "those unfortunate zealots who are to be respected for the goodness of their cause, and pitied for the injudiciousness of their advocacy," when they really commiserate us that as scientists and practitioners we have not the array of facts or the closeness of classified observation required to vindicate us in the use of alcohol as a food or in its indefinite prescription as a medicine.

Anstie, in one of his last articles, "Remarks on certain recent papers on the action of alcohol," presents the views of Ross, Bintz, Beale, and himself, and seems to concur in the view that "its chief therapeutic use" is that it interferes with blood fluidity and the disintegration of blood corpuscles and hardens the vascular walls (see "Practitioner," 1873, p. 361, etc.) And Ross elsewhere claims (*British Med. Journal*, p. 396) that it "checks growth and consolidates structure" and delays white corpuscles. These are but hypotheses, and if true, such results could not readily be accepted as processes necessarily limiting disease or assisting recovery.

Dr. Wm. H. Dickinson, in his thorough papers on the morbid effects of alcohol, after a series of tabulated facts as to various diseases, says: "Alcohol certainly gives an asthenic type to disease. Although we can not as yet say with certainty that it defibrinates the blood, yet it retards adhesive and plastic processes." He recognizes it also as making degenerative substitute for "vital material."

When Beale suggests that alcohol does good by diminishing the white blood corpuscles (see "Protoplasm," p. 244) we must read beside this the testimony of Dalton, that we have little determinate knowledge about the white corpuscle, or a recent paper on Leucocythæmia, which shows that we are not yet able to differentiate them from other corpuscles as factors in disease.

It will not do to say that alcohol "impedes the blood changes, by which oxygen quickens nutrition, saturating the blood corpuscles and liquor sanguinis," and then call that "recuperative delay of metamorphosis," unless you tell us what all this means, show how the facts are proved, and then show how such results are healthful. Nor can we generalize such a great disturber into respectability by such a platitude as that "it equalizes the circulation."

Dickinson (Vol. V. "Med. and Chi. Trans.," 1873), in his accurate paper, concludes his experimental and clinical research with this sentence as the summation of his experience:

"Alcohol replaces more actively vital materials by oil and fibrous tissue; it substitutes suppuration for new growth; it promotes caseous and earthy change·

it helps time to produce the effects of age, and, in a word, is the genius of degeneration."

Such a deliverance is certified by the immense fact that alcohol is the article that makes more sickness, disease, and death than any one article claimed to be used medicinally. Chambers says of it, that it causes " three-fourths of the chronic illnesses which the medical man has to treat."

Recent reviews and studies, chemical, physiological, and clinical, thrust it upon us when we think and act with all the exclusiveness of rigid professional scrutiny, and bid us review our therapeutics lest we stultify ourselves as physicians in our attempt to be conservative as citizens. It is asking too much of us to be empirical as doctors lest some medico-beverage advocate should stigmatize us as a profession of reformers. We are called upon to vindicate its use as a medicine by the most exhaustive evidence of its curative agency.

Any article in fidelity to our science and art has a right to claim exact analysis and close clinical testing, but this article also demands it with the additional importunity of imminent and vital interests pending thereupon and of deepest concern to all peoples. Nay, more, are we not called upon, in the relations which this article bears to the populace, if it is a possible medicine, to inquire how far it is a *necessary* medicine? So far as nutrition is concerned, we certainly have come to understand more thoroughly the laws of alimentation, and to see how much more can be accomplished by nutrients than by alcoholics. Where a stomachic effect is sought—ginger tea,

6*

beef tea, common tea and coffee, effervescent draughts, ether, or ammonia, will produce the local carminative or warning effect sought.*

In my own experience I have valued alcohol most as aiding to check that tendency to wrong fermentative actions in blood and tissue which occurs in erysipelas, typhoid fever, or low grades of zymotic diseases. As, having undergone fermentation, it might have a preservative or arresting effect on degraded tissues. But acetic acid, which has also undergone its fermentation, and to which, in the system and out of it, alcohols tend, will do the same. So also tinc. ferri chloridi, chlorals, potassium, quinine, ammonia, and other remedies are available. But since various other alcohols have come into use, we are not compelled to resort to such as are used for beverages. Advance in medical knowledge also enables us in feebleness of circulation to avail ourselves of position, of surface-medication, and of other restorative means.

Actual practice in syncope, or *coup de soliel*, and in fevers, does not reckon alcohol as indispensable as

---

* At a recent meeting of the Church of England Temperance Society at Oxford (October, 1876), Vicar Acworth related the incident, that once, when broken down from overwork, he consulted Dr. Conquest, of London, who prescribed for him port wine each day. "But," said the vicar, "I do not like the example." The doctor replied: "Oh, if you are inclined to take physic, I can give you physic that will answer the purpose equally well." "So he wrote a prescription, which I got made up for a shilling; and, at the end of a fortnight, he said I was all right. If I had begun to take the port wine, I should probably have been taking it to this hour."

it once did.   The rationale of treatment, and the results of available and successful methods, exclude
it or render it unnecessary.

Read, for instance, any one of the three or four
best sustained methods for. resuscitation in asphyxia,
for the treatment of sunstroke, for the management
of sudden syncope or spasm, and we can not but be
impressed with the very secondary assignment now
given to alcoholics.

The direction of Tanner in delirium tremens, " En-
force total abstinence at once," would scarcely have
come from a conservative authority thirty years ago.

In neither hyperæmia nor anæmia do we now look
to alcoholics for sovereignty.   " In both cases," says
Neimeyer, " the brain lacks its new supply of arte-
rial blood."   This is because the requisite oxygenated
blood-supply to the nerve filaments and ganglionic
cells is interfered with.   Alcohol has no *rôle* to play
in such an emergency, except as an arousing irritant,
within the narrowest and most precautionary limits.

If we are forced to look to the nervous system for
its therapeutical sphere, we are no longer left to rely
upon it.   The bromides, quinine, ergot, chloral, camphor bromide, caffeine, or nux vomica, the ethers, and
other articles give better-ascertained and better-sus-
tained results.   The assertion that ammonia is not so
good because it does not act on the cerebro-spinal system (Morris), is not in accord with the practice of
many clinicians who resort to it and to the ether-
wines when decided alcoholics fail.   Some of these
substitutions are not optional, but demanded on direct therapeutic grounds.

We believe the time has come when as advisers of this medicine only for strictly therapeutic and professional uses, we must seek from the chemists or pharmaceutists, like Squibb, or Powers & Weightman, etc., a uniform ethyl alcohol, just as we seek any other pure medicine. We must transfer our allegiance from the beverage-makers to such. If there is medicinal virtue in the alcohol, let us insure a reliability in kind and quantity, which is the start toward the clinical observation which we can tabulate into experience. We can then label it *medicine*, and estimate its effects.

If, on the other hand, the value as of wines is more in the ethers, or as of whiskies in the essential oils, etc., or as of brewed liquids in grain extracts and hops, let us have these for medicines in definite and ascertained proportions, or have those definite chemical imitations which it is now claimed are equally medicinal as the natural products.

If we would study clinically the effects of any remedy, we are not at liberty to use all alcoholics indiscriminately and talk of our experiences over unknown mixtures. Edward Smith, therefore, in his book on Foods, well insists that we must study liquors individually.

It is only by such discriminations that we shall make effectual advance toward the settlement of therapeutic value, and at the same time conform our inquiries to the requisitions of our art. Thus only can we conform to " the rigorous methods of modern science whose effort is to disengage itself more and more from any preconceived idea, and to confine itself to establishing absolute and constant relations between the facts and the antecedent conditions."

This paper has relied upon strictly professional authorities and professional experience. It has not attempted to define the mere attitude of physicians toward popular drinks.

But it avers and argues that certain beverages shall not be permitted to thrust themselves into the domain of therapeutics and pass themselves off as medicines without that scrutiny to which medicines are entitled, and without that exact analysis of result with which outside prejudice in their favor is now under suspicion of interference.

Let us draw close and taut the dividing line between what we use as a medicine and what others use as an acceptable beverage. Let us have such purity of article, such source of supply outside that of sumptuary indulgence as that we shall have a definiteness and exactness both of knowledge and of use, such as mere beverage considerations can neither make nor mar. Let us not as physicians be inveigled or beguiled into a tacit acknowledgment and actual support of an article as a beverage when our only relation to it is that of officinal prescription. The drug store " liquor-dealer" must not have our abettal.

If confining ourselves to its use only within the legitimate bounds of careful prescription, we are in our place invulnerable, but we must guard lest loose medical views and so-called medical use degenerate into unprofessional laxity, which gives to popular practices unfortunate momentum. We believe to-day more depends upon pronounced and discriminating views, on the part of our profession, in regard to this article of the Materia Medica, than upon all other reasonings from all other men combined. The true status of alcohol is to be decided at our forum. A defined and precise therapeutics would do much toward limiting its assumed right of eminent domain. But even outside of such consideration, and inside our own scientific art, we are compelled to inquire whether, when we give a general permit of use for an alcoholic, we are not forsaking the rigidity of medical prescription, the exactness of skilled diagnosis and medication, and really to be counted rather as public caterers than as strict therapeutists? Are we not often yielding, as an excellent recent Materia Medica suggests we should do, to the preference of the patient rather than prescribing with the exactness of the physician?

With all the dignified talk about having to do with alcoholics only as medicines, and having nothing to do with their beverage aspects, are we not too often beveraging our patient with an easy virtue of accommodation, and not primarily medicating him with an expert adaptation of remedy to disease?

In a case like this, not now by moral appeals, but by response to fair therapeutic requisitions—all the more pressing because of relative interests—we would draw so significantly the *cordon sanitaire* between the beverage and the medicine as to prevent such communication as loose prescription and loose therapeutics are sure to establish.

Thus we may come really to know whether purely medical considerations are controlling us, and whether a careful scrutiny of the article itself, of the disease, of the treatment instituted, will vindicate our prescription.

If such is found to be the case, then and thus will the advanced thinkers on social and moral questions grant that we are fully justified in our medicinal pre-scribings, as knowing that this is the work assigned us, and that we are respecting the safeguards of a distinct boundary.

If, perchance, as we opine, it shall, on the other hand, appear that alcoholics are needed as medicines only about as frequently as podophyllin, camphor, or colchicum, we shall be able to excuse ourselves from manifold suggestions for its employment. We shall thus only be as true to our art as we ought to be in-dependent of any collateral issues. It is high time, too, that we avail ourselves of riddance from respon-sibility for some prevalent evils, by drawing intelli-gently and distinctly the line between the beverage and the medicine. We can not conceal from our-selves as physicians, that thousands with sincerity indulge in the use of alcoholic stimuli because they entertain the idea that health requires it. Some physician had advised a little wine or brandy or ale for some ailment, and the patient continues the prescription, or renews it repeatedly, because "his constitution requires it." We have been sadden-ed to find those pledged to total abstinence thus using the beverage, and really deceiving themselves. So exceptional is the indication for the use of alco-holics in any chronic ailment, that no one who claims to be using his drink as a medicine should forget to consult his physician *very frequently* as to the need for its continuance. If such were the rule, and if physicians exercised the clinical precision as to it which is demanded, thousands who now say they use it medicinally, would cease its use or know they were, at their own risk, enjoying a luxury, imbibing a beverage, and sapping the vital forces by a toxic.

If to-day no physician would advise any patient to the use of any alcoholic drink, but restrict it within the close limits of his particular prescription, the limitation would be in harmony with the present demands of therapeutic knowledge. It is not merely that the morals of society would get a glorious health-lift, but the act would knock away the false prop which now upholds so many in the use of alcohol, and relieve us from the imputation of being accessory to the perverted habits of multitudes. If men and women will call it a food because they like it, they must cease to quote the medical profession as authority until there is proof that it has some ascertained value as such. If they wish to use it under the plea of medicine, and make self-prescription for their own gratification, they must not do it by our sanction. The facts as to food dismiss it as such. The facts as to medicine confine it within boundaries so narrow, that we must, in fealty to real science and right practice, hold it closely within its limits.

Wandering beyond these, it must in nowise identify us with its vagaries. Because it finds a place in our Therapeutics, it behooves the medical profession to locate and define it. This we have sought to do, and respectfully present and submit the following conclusions :

1. Alcohol is not shown to have a definite food value by any of the usual methods of chemical analysis or physiological investigation.

2. Its use as a medicine is chiefly that of a cardiac stimulant and often admits of substitution.

3. As a medicine it is not well fitted for self-prescription by the laity, and the medical profession is not accountable for such administration or for the enormous evils arising therefrom.

4. The purity of alcoholic liquors is in general not as well assured as that of articles used for medicine should be. The various mixtures when used as medicine should have definite and known composition and should not be interchanged promiscuously.

THE BIBLE WINE QUESTION.

The National Temperance Society has published a variety of Books and Tracts upon the Wine Question, by some of the ablest writers in the world. The investigation clearly shows the existence of two kinds of wine, the fermented and unfermented, and presents numerous and convincing authorities. The following is a list:—

Sent by mail, post paid, on receipt of price.

Address J. N. STEARNS, Publishing Agent, 58 Reade St., New York.

Talks on Temperance.

BY

REV. CANON FARRAR, D.D., F.R.S.

12mo, 198 pages; cloth, 60 cents; paper cover, 25 cents.

———— ••• ————

THE NATIONAL TEMPERANCE SOCIETY has recently published the **12** Sermons and Talks by this eminent divine. They are filled with sound convincing arguments against the lawfulness, morality, and necessity of the Liquor Traffic, as well as stirring appeals to all Christian men and women, to take a firm, decided, outspoken stand in favor of Total Abstinence from all intoxicating liquors.

He gives the trumpet no uncertain sound, when he proclaims war against Alchohol, but urges every motive, and brings to bear every incentive, to enlist recruits from every class.

OVER 40,000 COPIES

have already been sold in England, and we trust that, with the very low price at which they are sold, they will secure a wide circulation in every community. The following is the Table of

CONTENTS:

It will be sent by mail *on receipt of price.*

Address   **J. N. STEARNS, Publishing Agent,**

58 Reade Street, New York

www.ingramcontent.com/pod-product-compliance
Lightning Source LLC
Chambersburg PA
CBHW060748100426
42813CB00004B/743